LIBRARY
College of St. Scholastica
Duluth, Minnesota
WITHDRAWN

false profits

false profits

the decline of industrial creativity

Thomas P. Carney

UNIVERSITY OF NOTRE DAME PRESS
NOTRE DAME LONDON

Library of Congress Cataloging in Publication Data

Carney, Thomas P.
False profits.

Includes index.
1. Creative ability in business—United States.
I. Title. II. Title: Industrial creativity.
HD53.C37 658.5′7 81-50460
ISBN 0-268-00851-5 AACR2

Manufactured in the United States of America

contents

preface

I suppose I have been spoiled. I feel sorry for all those people who were not lucky enough to have been born at a time that allowed them to develop in an atmosphere where they had the opportunity to do everything—to be free to think, to be free to act. I feel sorry for the scientist today who must work in a narrow area of science, who must work on rediscovering yesterday's ideas, who must work in organizations whose leaders themselves do not feel free to lead but who react with Pavlovian predictability to signals given to them by a new breed of bell-ringers called "financial research analysts."

The technology to which I was exposed in the 1930s, 40s, and 50s produced miracles almost as a routine. Even in the 1920s products such as insulin and the liver extracts were made possible by the cooperation of industry and the academic community. These two products alone have saved millions of lives, and millions more, formerly condemned to a life of suffering, have been rescued to serve out their allotted lifespans as useful citizens. Some vaccines were developed in the 20s and early 30s. But it was in the late 30s and early 40s that the real time of miracles began.

My first job after I graduated from college—I was a chemical engineer then—was with a small company making chemicals from coal tar. It was a company that recognized the importance of being innovative, but it was also a company with few of what we now recognize as the technological resources necessary to have separate groups to take an idea through research to development to production and ultimately to sales. It was thus that, as a result of an entirely unplanned first step on a "career path"—it really was the only job available, and I had never heard of a career path—that I had my first experience in being able to do everything: to do research, to do development, to work in production, and even to work with salesmen.

Following some post-doctorate work I then joined a company whose record shows it to be one of the most creative in the world. Such products as insulin, liver extracts, the barbiturates, Merthiolate, the first biosynthetic penicillins, the cephalosporan antibiotics, and the first nonnarcotic analgesic were either originated or developed there. It was this laboratory that saved the Salk polio program from disaster.

It was a company that was willing to work, also, on basic or pure research. It was the first industrial laboratory to have a tissue culture program before there were any known uses for tissue culture. It first synthesized lysergic acid, and it was the first to isolate and determine the structure of a hormone, Glucagon, in its own laboratory. It extended its creativity into the agricultural field, and developed herbicides and growth stimulants, using the same quality of research as it had expended on drugs.

I then became associated with the laboratory that had originated the first birth control pill, and Probanthine, and Dramamine.

During thirty-five years of research I became familiar with the programs in all areas of technology. In 1970 I was one of a group of ten scientists selected by the National Academy of Science and our State Department to visit India to evaluate all fields of industrial research in that country at the request of the Indian government. I was a member of a government commission to consider the practicality of technology transfer to developing countries. I was a member of a commission to advise on the patent system in this country, and I was on the Bayne-Jones Committee, the committee to advise the Secretary of Health, Education and Welfare on the status of medical education in this country.

If, then, I speak about creativity, and value creativity, and evaluate creativity, and mourn for the passing of creativity, I think I know whereof I speak and value and evaluate and mourn.

In the following chapters I shall discuss my reasons for believing creativity has declined in industry. I shall talk about the environment within companies and how it has changed over the years and the reasons for the change. I shall discuss the external factors—the social pressure and government controls—that influence decision-making in corporations. And I shall discuss the reasons that lead me to believe that, although new products will probably be made available through industry, there will not be creative, major contributions originating in industry unless the basic work leading to these developments is already well along toward completion, or unless the attitudes of both government and industry leaders change.

1 . . . wherein business is good and bad, and business leaders are led more than they lead

At one point in Charles Dickens's *A Christmas Carol,* Scrooge says to the ghost of Jacob Marley: "But Jacob, you were always a good man of business." The ghost replies: "Business? All of mankind was my business. The common welfare was my business; charity, mercy, forbearance, and benevolence were all my business. The dealings of my trade were but a drop of water in the comprehensive ocean of my business."

In these days of consumerism, public activism, and social consciousness, I suppose every business executive would like to fit Jacob's self-picture—as long as profits could also be maximized.

Polybius, the Roman historian, gives another picture of a nation's business: "At Carthage, nothing that resulted in profit was considered disgraceful." Probably the real picture of the vast majority of businesses and executives is somewhere between these two extremes. Profit is necessary or there is no business. And, at least in recent years, most businesses have recognized a responsibility to the community in which they operate and to the people they serve and who serve them.

But what is business really like? What is business?

It is the greatest influence for good of any organization in the world, and, at the same time, has probably been responsible for more evil than any institution other than a corrupt government. It is stronger than most governments, yet is totally at the mercy of most governments. It is a faceless, cold, efficient organization whose objective is to make profit, yet it is absolutely necessary for the social progress of the world. It is one of the biggest users of government money, and the largest contributor of taxes to the government.

The extreme conclusion to be drawn from the above is of an institution of great power, but responsible to a number of different constituencies. First of all, the corporation—and now I am speaking of an individual corporation—is responsible to the owners who, in the case of a large company, are the shareholders. The company is responsible to its employees. It is responsible to its customers and to the vendors from whom it purchases materials. It is responsible to its government, and it is responsible to the general public. All of these responsibilities are important. Depending on the objective of the individual company, they might even be in conflict with each other. If the objective is to maximize profits, then the lower the wages and salaries that are paid, the higher the profits that result. If the company has a large charitable contributions budget, then the company will make less profit, but the company will appear to be more socially responsible. If a corporation recognizes tax havens to be used to its advantage, profits will increase, but contributions to government will decrease. It is these seeming conflicts that result in business being seen in so many different lights depending on the sympathies of the viewer.

It is the same apparent conflicts that make it difficult for a company to arrive at corporate goals and objectives. The management of the corporation would, with Jacob, like to appear to "have all of mankind" as its business. It would like to be known as the company that pays the highest wages and salaries in its industry. It would like to be the company cooperating totally with the government to produce social gains. However, the cold fact is that, in determining company goals, profit is the dominant factor, and everything else is compromised to that end. Profit-making must be dominant in decision making, but the extent, not the necessity, to which other things are sacrificed determines the character of a company.

Obviously, a company must make a profit to survive. In addition to the necessity to pay its bills, why does a company want to make profit, and why does it want to make a lot of profit? It wants to be profitable for two reasons, and I shall mention the most obvious one first. Large profits attract investors to the purchase of the stock which, in turn, results in the price of the stock increasing. Very little stock is purchased today for investment in the true sense of the word, that is, in order to get a high return—in this case a high dividend—based on the cost of stock. Rather, whether purchased by individuals or institutions, the objective is to hold it for a relatively short time, and then obtain a return from the profit of sale, rather than return of dividends. In this respect, then, most stock purchases are speculations. Thus the attraction for high profit in the company.

What makes a stock increase in price? It is simply supply and demand. The more people who want the stock, the more it will rise. The people want it because it rises, so where does the cycle begin?

The decision to buy stock is not made on as simple a basis as just increasing profits. There has developed over relatively recent times a group of people known as research analysts who wield a tremendous power over the performance of individual stocks. I suppose, overall, I would conclude that the average investor, particularly the small investor, but even some very large institutions, is influenced more by emotion than by cold, critical analysis. The fact that his or her goal is to make money on a rise or decline of the stock makes the investor particularly susceptible to reacting to any rumor of action that affects stock prices, and lack of expertise and knowledge contribute to this susceptibility. Therefore, opinions of experts are given great credence, and result in buying and selling.

The job of a research analyst is just what the title sounds like. He or she does research on the operation of a company, analyzes information, and then comes forth with a conclusion usually resulting in a recommendation to buy, sell, or hold a particular stock. Some even write newspaper columns. There is no single group who wields such influence over the price of a stock with such little justification. Ordinarily, the analyst has no more information about a company than does the stockholder or the general public. In fact, it is illegal for a company to reveal significant information to any one individual or group of individuals without making a general announcement of the fact. I can only repeat that the reaction to any pronouncement from the experts is based on emotion. How often have you seen a stock drop after a company has had a record report, with the drop accompanied by the explanation that the gain did not come up to the analysts' expectations, even though it was a record gain? How often have you heard of a stock price increase or decrease being explained simply by the fact that there was a favorable or unfavorable mention in *The Wall Street Journal*, or *Barron's* or *Forbes* or *Dun's*? In particular, I have analyzed one of the daily columns in *The Wall Street Journal* over several years. The column quotes analysts on various stocks or gives the analysis of the columnist.

Almost 100 percent of the time the stock, on the day of the report, rises or falls depending on the recommendation. Characteristically, if two analysts make opposing recommendations, the stock falls. The pronouncements of the analysts, therefore, assume the guise of self-fulfilling prophecies.

The following are just some of the hundreds of examples, all taken from *The Wall Street Journal.*

October 20, 1977—Merck, which produces an anti-arthritic drug, fell 2 to 52¼; a *Wall Street Journal* story noted analyst views that doctors are going back to aspirin in the treatment of arthritis.

October 28, 1977—General Motors was third most active stock on the big board and fell ½ to 67½; some analyst said earnings didn't improve in line with the record third quarter sales.

November 18, 1977—Atlantic Richfield slid 1¾ to 50¼; all *Wall Street Journal* stories noted lower analysts' estimates for the stock.

November 22, 1977—General Foods, mentioned favorably by analysts quoted in *The Wall Street Journal* story, rose ⅝ to 33⅛. (And from the same column on the same day) General Motors, in third place, slipped ½ to 65¾; a brokerage house downgraded the stock.

December 16, 1977—Kirsch jumped 1⅜ to 24; a *Wall Street Journal* story discussed analysts' favorable expectations about home furnishing stocks.

January 17, 1978—Cummins Engine, which received unfavorable analyst comment in a *Wall Street Journal* article, slipped 1½ to 34.

January 24, 1978—On the AMEX Data Terminal Systems tumbled 10½ to 59; a *Barron's* story reported on the growing competition in the electronic cash register business.

January 27, 1978—The glamour issues were hit particularly hard. McDonalds, downgraded by a brokerage firm, slipped 1½ to 44½ in active trading. Digital Equipment, also active, fell 1½ to 41; a *Wall Street Journal* story noted negative analyst comment about the stock.

March 24, 1978—Caterpillar Tractor slipped 1⅛ to 45⅝, and Merck 1¼ to 49½. A *Wall Street Journal* story noted negative analyst comment on the stock.

July 20, 1978—Weyerhauser advanced 1⅜ to 27; a brokerage house raised its earnings estimate for the stock.

On October 18, 1978, the *Journal* reported the following: "Shares of Memorex . . . were delayed in opening and then knocked down as much as 11⅜ points, or 23%, at one point in heavy trading before the stock closed at 38¾, off 10⅜." What sort of catastrophe took place to justify such reaction? How much did the company lose? Could it survive?

The catastrophe was that the company earnings did not reach the level that analysts had predicted for it. The company did not lose money. It increased its earnings to $1.30 per share from $1.27 a year earlier. To quote the *Journal* again: "But profit was at least 30¢ a share below estimates being used by some Wall Street analysts and also was under the $1.45 per shares earned in the second quarter . . . . And although revenue in the quarter rose smartly to 152.6 million from 114.1 million a year earlier, it still was as much as 20 million short of what some analysts expected." The important thing was not what the company had done, namely, increase its profits, but that it hadn't met the expectations of a group of outside analysts.

On April 12, 1979, headlines in *The Wall Street Journal* announcing IBM first quarter results said: EARNINGS OF IBM ROSE BUT ONLY 13% IN FIRST QUARTER. Analysts had expected more. The stock skidded $8 a share to $312. The business section of *The Chicago Tribune* reported Sears first quarter 1979 earnings with the headline: SEARS EARNINGS DIP—BUT LESS THAN EXPECTED. The stock rose from $19 to $19½ on the day of the announcement. IBM earnings rose, but the price of the stock went down. Sears earnings fell, but the price of the stock increased. The performance of the companies seemed almost incidental to the importance given by the investors to the expectations of the analysts.

Polaroid experienced similar difficulty when it reported its earnings per share for the first quarter of 1979 up "only" 18% over the previous year's quarter. Analysts had expected more. *The Wall Street Journal* headed the article ANALYSTS SEEK POLAROID DATA IN MEETING IN WAKE OF DISAPPOINTING FIRST QUARTER RESULTS. Polaroid also ran into difficulty with the analysts from another direction. An article of March 1, 1979 headed, POLAROID'S RECENT PRICE DROP IS TIED BY ANALYST TO COMPANY'S POLICY OF SECRECY, stated: "While other companies whose prospects also depend on glamorous new products . . . talk regularly with brokerage house analysts and institutional investors, analysts complain that Polaroid as a rule turns off requests for more information, discourages visits and rarely makes presentations." A Polaroid vice president explained his position: "Our product cycle is measured in years, as long as a decade.

Because our financial results in short spans of time don't exactly fit the dimensions of our major product cycles, we let the results speak for themselves."

Possibly the most dramatic example of the influence of the recommendation of a single individual on the stock-trading public occurred on January 7, 1981. In this case the individual involved operated a financial advisory service through subscription to a Market Letter. For $250 per year, more than 10,000 subscribers received, by means of the Letter, advice on when and what to buy or sell. Additionally, for $500 per year, more than 2,000 subscribers became the beneficiaries of an "early warning system." This warning service included individual contacts from the service in case of significant and sudden changes in the market.

On January 6, 1981, the Dow Jones average had passed the 1,000 mark for the first time in four years. Only two days before this the Market Letter of the financial advisor had given strong advice to buy—buy anything—because "regardless of what you buy right now it should be higher several weeks from now." However, the evening of January 6th the advisor changed his mind, and the early warning system went into effect. As the advisor described the procedure in sixty interviews that he said he granted in the next two days, 30 operators, beginning at 6:30 in the evening, began advising the early warning subscribers by phone or telegram to sell everything they owned. By 2:30 that morning the job was done. And sell the clients did. The markets opened to a flood of sell orders. By the time the markets closed on that day, the Dow Jones average had fallen 23.8 points. Sales of 92.89 million shares on the New York Stock Exchange were the highest in history, eclipsing by more than 8.5 million shares what had been the former highest day's sales. The loss in value of all the shares traded on United States markets was about $40 billion for that single day. The slide continued into a second day, when the Dow Jones dropped 15.19 points on sales of 55.35 million shares. By the third day the Market Letter lemmings had fulfilled their destiny, all their stock had been sold, and the market returned to normal, closing the day with a gain of 2.99 points for the Dow Jones average.

Certainly I do not impute any evil motives to research analysts or financial advisors. Their actions are not conspiratorial, or subversive, or illegal, or immoral, or unethical. I admit I don't understand how someone could follow blindly and obediently advice given by a voice in the night that says: "Go, sell all you have." Such advice wasn't even followed in the bible. Possibly the reason for the difference in reaction to the biblical advice and that given by the present advisers is that, although the voice giving the admonition in the bible was no less powerful than the voices presently giving financial advice (some advisors would dispute that) the motives for selling were different.

I don't think the prediction of the analyst can make an inept company successful, or make a successful company fail. However, the price of a stock is an important factor in the strategy of a company, for example, in making acquisitions or going into the market to borrow money. Therefore, even a temporary change in the value or direction of change in the value can affect the strategy of the operation of the company, so it is important for a company to make decisions whose effect will be an increase in stock price.

Earlier I said there were two reasons why a company's objective was to make money, and said that the attraction of investors was one. The second reason is harder to perceive, and even harder to quantify. Simply stated it is that the management of any company wants to look good in the eyes of their peers, the business community, their neighbors, and the general public. It is a curious thing but, in modern times, individual managers of companies are becoming less well known as individuals. In the past, the old-time, rugged individualists, the entrepreneurs, the risk takers, the innovators, stood out. In these days of committee management, or collaborative management, individuals have a more difficult time being recognized for their specific contributions. Awards are made to management, not to individual managers. I venture to say that it would be difficult for five out of 100 people to name the leaders of ten of the Fortune 500 companies in this country.

In spite of that, each individual manager knows who he or she is, and believes everyone else knows. Therefore, when a management is declared to be good or bad, the manager assumes it is a personal tribute or adverse criticism, and reacts accordingly. It is a natural reaction.

If this is true, I think it is obvious that objectives of companies might reflect the personal egos of the top management. Management in most corporations is judged by the amount of profit their company makes, by the steady growth of earnings from year to year, and by the rate of success in the risks it takes or, to put it another way, by the number of failures it has. Although individual managers might not be known widely as individuals, within any company top management means one or two or three or a small group of people. They know they are management. Therefore, they react as individuals. The way to obtain acclaim is to meet the expectations of those responsible for influencing and making such judgments. As mentioned earlier, investors buy a stock for its anticipated increase in price. By the very nature then of the factors determining stock price, the decisions affecting the short-term operations of the company inevitably take precedence over those resulting in long-term benefits, since it is today that judgments are being made.

Executives, then, whose egos must be satisfied by the approval of their peers, and who want to be as sure as possible that they will succeed in those things that bring approval will invariably eliminate as much risk as possible from their decisions. This, also invariably, means that only those activities that will bring a fast return, certainly no longer than a year into the future, will be undertaken.

I suppose it is natural that managers react in this way. It is, in effect, reacting to public opinion. It is impossible to determine whether or not such decisions actually hurt a company. If a company is going along making a profit year after year, it is difficult to say: "yes, but if something had been done differently a larger profit would have been made." It is possible to say that such decisions certainly do affect the character of a company. Profits can be made as the result of products that result from long-range or short-range projects, or from products that result from high or low technology or no technology at all. But a company based on creativity, on producing products as the result of breakthrough technology, certainly has a different character and, for me, a more desirable character. I sometimes wonder what would happen if, by some magical process, the management of a company could be made truly anonymous so that no one would be given either credit or blame for results, that the managers would have to satisfy only their own needs for achievement. I believe there would be more risk taking and more decisions related to long-range benefits. I believe there would be more, rather than less, response to social needs. It would be nice, but impossible, to operate in an atmosphere where nobody cares who receives credit for progress. It would certainly result in decisions being made for the benefit of the company without regard to short-term judgments.

I think part of the difficulty arises from the belief that companies should be run by "professional managers." Certainly no one would want an "unprofessional" responsible for a company. But the word "professional" now has a very special connotation. It means someone who has been trained to direct a business. Business experience is not sufficient. Some exposure to a formal business education in one of the business schools is the passport to the fast track up the supervisory ladder.

Unfortunately, as with most training, the resulting product is a carbon copy of everyone receiving the same training. There are certain rules and regulations learned that are necessary to operate a business, and everyone learns the same rules and regulations. Almost by definition the professional manager does not take a chance and does not do anything really new. The result is usually someone who does a competent job in operating something that is already in existence, possibly doing it better and maybe even making it bigger and more profitable. But the very rules that specify that a company will be run safely usually make it impossible for the company to expand. The manager can *operate* the company, but he or she cannot *build* a company.

The other unfortunate side effect of this kind of professional criteria is that the resulting leaders seem to be almost entirely absorbed with the day to day internal operations of their companies to the exclusion of any external activities. Most company leaders would prefer not to be involved in public relations, in public disputes, in defending themselves against attack. Maybe some even think they shouldn't expose themselves to these indignities. But a large company, or even a small company, is a force for social change and influence these days. Presidents of companies are on a par with public officials. They should be representing their companies in the public forum just as a member of congress or senators or public service activists represent their positions.

In addition to the fact that many leaders of industry don't recognize their responsibility to be public figures is the element of fear, fear of being misquoted, fear of "looking bad," fear of bad public relations. But the alternative to speaking up is to lose by default. So the leader falls back on professionalism. He or she is doing a job by running a profitable company, making the profits grow just as they have been trained to do and just as the rules say they should.

What should be considered adequate profits for a business? I do not know, nor do I think it can be determined quantitatively, nor do I think the same standards can be applied to all companies. A company operating in a high technology, high risk industry, where a high percentage of sales dollars must be expended in research, requires a high percentage profit on its investment to encourage it to take risks. A low risk business requires a lower percentage of profits. There can be at least one general rule—every company must make a profit over time or it cannot survive.

My concern is not with how much profit is made. It is the method by which companies arrive at their decision to make a profit and, critically, how these methods affect decisions that will either increase or decrease creativity.

I think the major compromise that companies make to meet the expectations of the financial community is in trying to fulfill the expectation of continued growth. It is not satisfactory, in the eyes of those recommending investments, to grow at a rate of 10 percent increase for one year. Now it is required to grow at the same rate every year. If, in the fourth or fifth year, growth slows to 6 percent or 8 percent, this is an immediate signal to dump the stock. I think it is ridiculous, since it is possible that percentage growth has decreased as a result of increased spending for even more rapid future growth. Regardless of the facts, it is the accepted norm that a company, once having established a growth pattern, must not fall below that pattern for any reason.

There is no convincing argument these days for the necessity for a constant quarter after quarter high growth rate. But there is a reason—the reason I gave earlier. To the financial analyst, growth automatically means efficiency, good management, and therefore the price of the stock goes up.

There are several ways by which a company can grow. It can do more of what it is doing, it can do something different such as introducing a new product line, or it can acquire an existing company. Acquisition of an existing company is by far the most dramatic method of growth. A parent can double in size overnight with a single acquisition. But such growth may or may not be a reflection of a company's efficiency. There are two reasons for a company to seek an acquisition. The first is, obviously, to buy an operation that is already profitable so that the total company profits will increase. The second reason, and one equally important for a creative organization, is to enlarge the scope of its activities, in other words, to allow it to do different things and to take advantage of different ideas. The hallmark of the conglomerate is that it does control a number of different companies in unrelated fields. The acquired companies are operated as individual, independent units. The conglomerate, then, is the sum of the bottom line of these units.

This is not the type of growth I believe is desirable, particularly in companies based on technology. A company composed of a number of subsidiaries should not be characterized by the sum of its individual activities. Rather, it should be shaped by the interchange of ideas between units so that the final result will not be a summation but a synergism.

Too often when a company is considering an acquisition a decision will be based on how the acquisition will "fit in" with the parent. "Fit in" is another one of those catch phrases that means different things to different people. In many—probably most—cases, the desire is to have something so compatible with an ongoing operation that no great change is necessary. It should have a similar product line and the same customers so that the already established sales force can handle it. The rationale is "we don't want to expand into anything we don't know about." The more imaginative response might be "this is outside our present operation, but can we learn about it?"

The fear of something new, something "we don't know about," is not limited to the acquisition of companies. It applies equally to the acquisition of a product. I have seen many more products rejected because someone said "this doesn't fit into our product line" or asked the question, almost rhetorically, "how can we sell it?" than I have seen rejected because the product was not good.

The textbook-trained managers appreciate, with good reason, the importance of knowing the customer for each different market. They know the importance of having a sales force already operating in a marketplace. But nothing in the textbook says that, knowing these things, you can't enter a market in which you are not now represented if you have a good product, if you study the market, and if you develop a sales force. The real value of an acquisition comes not in doing more of the things you are doing now or even doing them better. It comes in being able to do more different things better, and in being able to take advantage of the knowledge and ideas coming from a broader base.

Many companies have failed in attempts to diversify. In some cases the textbooks were right—the acquired company could not be assimilated because the operations were radically different. However, I think the majority of acquisitions fail because no attempt is made to discover the two-way street of contribution—not just from the parent to the acquired company, but between all divisions of a company back and forth and up and down. If an acquired company does not become profitable it is usually spun-off with the explanation that it was a poor choice to begin with. Possibly it never had a chance to succeed in the atmosphere created for it. Maybe, instead of saying "the company was a dog," it should be said "we loused up the operation."

But does growth really mean good management? It has become customary now to look at a company's earnings in terms of its operating profit, not its total profit. Operating profit is called hard, or quality, profit. It can't be influenced by a reduction in tax rate, for example, or by changes in accounting practices that allow credit for previously taken reserves. It can, however, easily be influenced directly by cutting down expenses. Where do expenses come from? Usually from people. Therefore, one way to maintain a growth rate is to eliminate those employees who do not contribute directly either to production or sales. Totally aside from the relationship of management to employees, it is just poor business. I am convinced that almost any company could be tremendously more profitable for a year or two if it were not interested in longer-term growth and expansion. This is particularly true of a company that is in a high technology area requiring a high expenditure for research and development. In very few companies do the results of the expenditure of research dollars show up in the profits for at least two years and, in some cases, the gap is as long as eight or ten years. The elimination of these research dollars, then, would be directly translatable to profits in that year, and there would be absolutely no negative effects on the company for whatever the time-gap between starting research and profiting from a product would be. The same is true for every other staff function of a company. Only production and sales would show an immediate effect if eliminated. Therefore, a company interested in increasing only this year's profits could eliminate planning, the personnel function, even the financial function. Such activities would almost guarantee the future demise of the company, but the example does indicate what the possibilities are for taking action to look good. I wish I could say it is only a hypothetical example. It is sad to relate that many companies today are, to a greater or lesser extent, taking these actions.

Companies listed on the stock exchange are required to publish quarterly statements of earnings, and many other companies do so voluntarily. Consequently, judgments of a company's progress are made, not just from year to year but from quarter to quarter. It is difficult to imagine progress that would result in profit being made from quarter to quarter in long-term research programs. It follows, then, that, if results that improve earnings are not obtainable, the expenditure for the project, whatever it might be, is not justified.

We have come to look upon Japan as our strongest competitor in the technological industries. In addition to the technological skills perhaps we should also consider their management skills. In reporting on the results of a Japanese acquisition of an American television company, and in analyzing the consequences, *The Wall Street Journal* made a significant observation. After reporting that the company was profitable, that after just four years it had increased its share of the market, and after quoting its employees as saying that working conditions had improved, the *Journal* said: "One early conclusion is that foreigners can be more patient than Americans about earning a return on their investment." They quoted Malcolm Salter, Professor of Business Administration at the Harvard Business School: "They are less inclined than Americans to want results now, and they aren't obsessed with quarterly earning statements."[1]

From the point of view of creativity, just as undesirable as eliminating innovative programs is the practice of simply not starting new programs. Departing from an established method of operation seems, in some companies, to be equated with taking a new risk. In addition the change in a production method usually costs money. Therefore, the tendency is to stick as long as possible with what is definitely known. This is true even when a change would result in improvement of a product or would have greater sales potential.

It is impossible to know for sure when a company has made a decision to maintain the status quo, even when it could increase the quality of its product by taking advantage of a change in technology. However, there are occasions when one must be suspicious of a company's activities. One such incident is illustrated by what happened in the razor blade business several years ago. The American companies dominating the market were satisfied with their share, and so were not interested in change. However, Wilkinson Sword, Ltd., of England, a company originally established to make ceremonial swords but now also producing garden tools, introduced a stainless steel razor blade into the American market. Because it was a superior product it very soon began to eat into the market shares of the established companies. The reaction was immediate. In an amazingly short time the American companies introduced their own versions of the stainless steel blade. There is no way of knowing, but one must suspect that the product had already been developed, but that without competition there had been no reason to change.

In most cases the company content to sit on its monopoly is not as fortunate as the razor blade companies who had been farsighted enough to plan for competition, if not progressive enough to take the lead in changing technology. A poem by Saki illustrates what usually happens:

> Some laud a life of mild content;
> Content may fall as well as pride.
> The frog who hugged his lowly ditch
> Was much disgruntled when it dried.[2]

In the days when single entrepreneurs ran companies and were responsible for the building of the company, a corporate character was the reflection of the character of those individuals. The company was their showcase. It was the stage on which each was the principal actor. If anyone else got into the act at all it was as a spear carrier or scene changer. Not many people use the term "character" when describing a company today. Now the word is "image." It is not, for me, a satisfactory word. Too often it implies what a company appears to be, not what it is.

The extent to which a company will go to preserve its image knows no bounds. Every company believes it is the best in the world. Every company believes its people are the best people. No company believes it has never made a mistake, but few companies will admit to errors, and every company will do its best to conceal them—not condoning, but concealing. I have never had any experience with the top management of a company who would willfully break the law, nor do I know of any company that did not act immediately to correct any transgression discovered in any other level of management. But wherever possible, every effort is made to avoid publicity. So sensitive is the image problem that managers, guilty of illegal or unethical conduct, have been retired with liberal pensions, rather than taking a chance on the publicity that might result from their firing.

Another area of sensitivity is the movement of executives from one company to another. Obviously, the company acquiring the executive sees it as a very positive action. However, a company losing a man or woman seems to take it as a personal insult that anyone could believe a different position would be more desirable. All kinds of rationalizations are made to explain the loss of top people. The expression, familiar to everyone in industry, that "we have never lost a good man," illustrates perfectly the attitude I am talking about.

I experienced this once in a very emphatic way. After having worked for a company for twenty years, ten of them in the capacity of vice president, member of the executive committee, and member of the board of directors, I decided to leave. I had, I think, made major contributions to the company during its growth both at a policy level and at an operating level. The parting was amicable and cordial. To this day relations are still friendly.

Several years after my departure, a history of the company was written to celebrate its hundredth year of existence. It was an excellent production, professionally written, and published in a hard cover edition as a book. My name was not even mentioned as having held a position in the company. My hurt feelings were somewhat assuaged when, a little later, I was presented by some of my former colleagues with a copy of the history, suitably and irreverently annotated by hand to commemorate the fact that I had at least occupied a space there.

Today a company is more than a platform for a single actor. It is more than just a place where individuals make enough money to allow them to live. Now the company is a stage where all the employees play out their parts. It is not just the place where ambitions are fulfilled, but where individuals begin to realize it is possible to *have* ambitions. The character of a company, then, is determined by its management and by its employees. It is determined by what products it makes or what services it performs. And it is determined by the policies guiding the activities, and the values guiding those establishing the policies. The company is no longer an institution whose only objective is to make money. It is an instrument that must also allow for the growth and development of individuals. If companies in the past were reflections of their founders and leaders, companies of today are also partially reflections of all the participants of company activity.

Probably the greatest quality that makes a company stand out from others, that gives it a definable character, is creativity. To the extent that a company is creative, to that same extent will it fulfill all the requirements now demanded of business as an institution effecting great social changes.

There are a few such companies left, but if there were an Endangered Species Act for companies they would certainly qualify for protection.

2 . . . wherein business becomes socially conscious, and social problems are hard to define, and the public wants a voice in defining them

In his book, *The Dragon*, Yevgeny Zamiatin says: "When a flaming seething sphere grows cold, the fiery molten rock becomes covered with dogma—with a hard, ossified, immovable crust. In science, religion, social life, and art, dogmatization is the entropy of thought; what has been dogmatized no longer inflames, it is merely warm—and soon it is to be cold. The Sermon on the Mount, delivered beneath the scorching sun to outstretched arms and rending sobs, gives way to slumberous prayer in some well-appointed abbey. Galileo's tragic 'E pur si muove' gives way to calm calculations in some well-heated office in an observatory. On the Galileos, the epigones build—slowly, coral upon coral, forming a reef: this is the path of evolution. Till one day a new heresy explodes and blows up the dogma's crust, together with all the ever-so-stable, rocklike structures, that had been erected on it."[1]

At the turn of the century, an industrial Galileo discovered that our standard of living could be improved by converting to an industrial society from an agrarian society. The flaming sphere of that objective grew cold, and the hard crust of profit dogmas covered it. The epigones built slowly the idea of profit, until the entire concept that evolved is the principle that business existed only to make a profit. Now, a new heresy says maybe business should do more.

A scientific Galileo once thought that science also could be used to advance the quality of humanity. The epigones slowly formed a reef that says science of itself is enough—it is neutral, it is amoral. A new heresy says that maybe scientists should assume responsibility for the results of science.

A dogma of perpetual growth was created. A new heresy says maybe growth is neither good nor necessary.

Heresies are not accepted easily. A revolution in business

practice does not take place without trauma, even when the revolutionaries are within the business community. When the impetus for change comes from non-business people the resistance is even greater. And when a change, regardless of its origin, involves the spending of large amounts of money for nonproductive activities, objections can be magnified.

Business has always had a profound effect on the sociological status of individuals or groups or entire societies. Mostly these have been nondirected effects, almost incidental to whatever happened as the result of the business operation. Along with other revolutions of the 1960s the attitude developed and pressures were exerted to see that business did consciously take a more positive approach to social actions, that is, that it developed a social conscience.

The expectations of business participation in social programs may be classified into three general types of activities. The first includes those things involved in the responsible and ethical operation of a business. The second involves those outside factors that are affected by the business itself, such as the physical environment, and the social and cultural environment of the community in which the business is located. Finally, there are those activities that might be described as contributing to national goals, such as urban development.

Most of the issues involved in the first class, namely, in running a business, are obvious and agreed to by everyone. A company should produce a quality product or service for sale at a non-profiteering price. It should abide by the laws of the nation and the community. It should have equitable personnel policies, including affirmative action programs opening equal opportunity to all. It should provide adequate safety measures for the protection of all workers.

Responsibility for hiring minority workers has been one of the most difficult for business to fulfill. Regulations state that each company must, as a goal, have in its work force a percentage of minority groups (Black, Hispanic, American Indian, etc.) equal to the percentage of the same groups representative of the community in which the business operates. Further, the distribution is not company-wide, but must be reflected in each major department. Further than that, representation must be proportional at various administrative levels. As an example, if a company is operating in a community where 20 percent of the population is black, the company is expected to have 20 percent of its employees represented by this minority. If the company has three distinct divisions, each division should also have 20 percent Blacks. Further, if in each division there are five departments, the same percentage distribution is expected. In each separate unit, whether it be a division or a department or the entire company, Blacks should also be represented in various levels of administration by the same percentages.

If a position is open, and if two equally qualified candidates are available, it is easy to select the minority candidate to fill the job. However, if the minority candidate is not qualified, the company is in a double bind. If the job is filled by an unqualified employee, it costs the company money, even though it does help fulfill the requirement of equal employment. The situation can be magnified if, in times of economic hardship, the company is faced with laying off an employee more qualified simply because he or she was not in the minority category. The specter of accusations of reverse discrimination is now being raised, and court tests are inevitable. A challenge in another field has already taken place in the case of Marco DeFunis, a white student who took his case to the Supreme Court because he felt that less qualified minority applicants had been accepted in a law school. The court declared the case moot since, in the time between the beginning of the suit and the hearing, DeFunis had received his degree from another school. In a second similar case the Supreme Court decided that Allan Bakke, a student who had been rejected by the University of California Medical School, had been discriminated against because less qualified minority students had been admitted. The Court ordered his admission. However, the decision was based on the fact that the school had a specific quota to be filled by minorities, and not that race was taken into account as one of the factors to be considered in evaluating the qualification of applicants.

In addition to those activities related to the operation of a business, much of the public expects corporations to engage in activities approaching philanthropy. Corporations are expected to contribute to charitable organizations, to the cultural activities of their community, and to educational institutions.

It was only in 1953 that a court test put the stamp of approval on charitable giving with stockholder money. The Supreme Court of New Jersey, in the case of *A. P. Smith* vs. *Barlow*, upheld a $1,500 corporate donation to Christian University. The court at least implied that business did have a social responsibility to higher education. Since then, other cases coming before the courts have uniformly been decided in favor of corporate giving. Since then, too, practically all corporations are on record as contributing to health, education, and welfare. Over 1,500 corporations have established foundations to make more professional the selection of recipients and the distribution of funds. The tax system recognizes the validity of such gifts by allowing a company to deduct up to 5 percent of its taxable income for such purposes. However, very few companies reach that level of giving. In an annual survey taken by the Conference Board, a non profit business research organization, 773 companies responded with details of their corporate giving. They reported company donations of $519.7 million in 1976, up from $436.8 million in 1975. However, this represented only 0.61% of their pre-tax earnings in 1976, down from 0.7% in 1975. Throughout the 70s the share has been declining. Health and welfare donations accounted for 39.1% of the total in 1976, with education being the recipient of 36.5%.[2]

In 1978 U.S. corporations contributed a total of $2.07 billion to charity, education, health, and welfare. This rose to $2.3 billion in 1979. In 1936, the first year that such information was recorded, total business philanthropy totalled $30 million. Obviously this is a tremendous increase and probably reflects not only increasing ability to contribute because of increasing profits but also an increasing sensibility of business to charity as a corporate responsibility. However, A. W. Clausen, president and chief executive of BankAmerica Corporation gives some figures that indicate business could do more. Of the approximately 2.1 million United States corporations "only about 25 percent make cash contributions and only about 6 percent contribute more than $500 per year."[3] In fact nearly half of all corporate philanthropic gifts come from fewer than 1000 companies. Stated another way, less than 0.07% percent of all corporations give one half of all corporate philanthropic contributions. There is by no means unanimous agreement that corporations should participate in any activity not directly related to the operation of a business and the making of a profit. In the opinion of some, philanthropy is an unwarranted use of stockholders' money. Every dollar spent in this area is a dollar lost to profits. One of the most distinguished and most critical of the opponents of this type of social responsibility is Milton Friedman, University of Chicago economist. In his book, *Capitalism and Freedom,* he calls the idea of a socially responsible conscience a "fundamentally subversive doctrine" in a free society, and says that such a society, "there is one and only one social responsibility of business—to use its resources and engage in activities designed to increase its profits so long as it stays within the rules of the game, which is to say, engages in open and free competition without deception or fraud."[4]

There have been some spectacular failures resulting from attempts by corporations to further social goals. When these occur, the chorus of protests against risking shareholders' money rises in intensity. For example, in 1971, Boise-Cascade found it necessary to report a huge loss because one of their efforts failed. In an attempt to encourage a Black-owned company in the heavy construction business, Boise-Cascade decided to guarantee the investment of that company. When it became apparent that the venture could not succeed, Boise-Cascade decided it could do no more and took the loss.

Very few socially significant programs have reached that order of magnitude in single areas. In the case of Boise-Cascade, it was an investment. Had the venture succeeded, the enterprise could have made a substantial profit while, at the same time, encouraging a minority business.

With the recognition by most managements that some form of social activity is a part of business responsibility, and with the demand of activist groups that all companies engage in this activity, has come the problem of evaluating the results of the social programs initiated. Two values are necessary—first, the return, if any, to the company, and second, the value to society or some specific segment of society. Industry has one objective measurement to determine its efficiency in the classical sense, namely, profit. Quantitative numbers are easily available. However, in the social sphere, it is not so easy to measure results with any degree of objectivity. It has been suggested that companies, in their annual reports, include a "social audit." The term is now well-recognized, but the fulfillment of the objective is far from being accomplished. If the "social audit" is to be comparable to the financial audit, then both company costs and company and social benefits must be measurable.

Possibly it would be more practical and sensible to settle for a social report rather than a social audit. The word "audit" carries with it an indication of quantitation and certification of the quantitation that is greater than it could be. Inevitably, there would be differences between the value placed by a company on a particular action, and that given to it by activists demanding more social action. A report of everything in the company considered socially responsible would at least give those interested a chance to make their own judgments. However, in an atmosphere where almost any business activity is suspect, where, with the passage of time, business is being expected to take more responsibility for social action, it is difficult to get agreement on what really is a business contribution. For example, if a company twenty years ago had a program designed to hire women and minorities in numbers that corresponded to the distribution of these classes in the area of the company's operation, it would have been classed as a very responsible action. With the installation of the Federal affirmative action program for Equal Employment Opportunity, every company that has a government contract of any kind is now forced to take this action. The program is rigidly enforced, with government inspectors demanding frequent accounting of progress. Therefore, since the company is *forced* to hire minority groups, it is no longer classed as a socially responsible action. On one extreme, to be classed as socially responsible, an action has to have some philanthropic aura about it. As one frustrated manager said in answer to what he thought was a lack of appreciation of his company's efforts, "If we don't lose money on it, a program is not socially responsible."

If we define a social problem as the difference between the social expectations of a group and the extent to which the expectations are being met by society, it is easy to see how that characterization of any single problem could change over time. I have already given one example—the hiring of minorities. If the expectations of any group are not high—for example, some classes of migrant workers—there is no social problem. There is, to be sure, a moral problem, but, under my definition, not a social problem.

Tocqueville, writing almost 150 years ago in France, anticipated our situation and explained it well: "The evil which was suffered patiently as inevitable seems unendurable as soon as the idea of escaping from it crosses men's minds. All the abuses then removed call attention to those that remain, and they now appear more galling. The evil, it is true, has become less, but sensibility to it has become more acute."

So we are seeing today agitation to change situations that were accepted in the past as normal. Sex and racial discrimination, as we describe them today, were not, within very recent times, even considered to be problems. Why? Because at that time the idea of escaping from discrimination "had not crossed their minds." Now, even though the situation has changed considerably for the better, the problem seems even greater because "sensibility to it has become more acute." The situation is improving not because the discriminators sought to solve the problem, but because those discriminated against finally recognized that they themselves could do something about it.

One of the primary jobs of business managers is to anticipate change, to prepare for it, and, if possible, to manage it. In this context, then, corporations should prepare to manage public expectations. One obvious and ideal way to do this is to operate in such a way that there is no gap—or at least as little as practical—between expectation and social reality, and still successfully operate a company.

But how can a company do this unless it knows what a social issue is? I have given an academic definition, and I have said that the same conditions in one period of time might not be a social problem, while in another they would be an acute problem. If my academic definition is acceptable—a gap between expectations and reality—what, then, are some of the expectations of individuals that are not being fulfilled? Here I find it difficult to specify any real goals for corporations that should not also be considered national or world goals. Probably the overriding expectation of all people is the hope that they will rise from poverty not, necessarly, to be rich, but to escape the consequences of being poverty stricken.

This applies to all peoples, so a corollary to that expectation is that it be made universal that each person have the same opportunities for developing himself or herself regardless of race, sex, or creed. Among the opportunities implied in such freedom are the opportunity to be educated, the opportunity to have a decent place to live, the opportunity for health care, the opportunity to have a good job or, in summary, the opportunity to be an individual equal to every other individual in dignity if not in the possession of tangible goods.

Obviously, business cannot be expected to solve all the problems. The public at large, however, thinks greater contributions should be made. According to many surveys, well over half of the public think that business has been remiss in its responsibilities.

In spite of the tangible benefits that have resulted from business of the past decades, businesses and business executives are held in lower repute today than at any other time in modern history. Perhaps it is just that critics are now more vocal. In any case, the prestige of business is decreasing, rather than increasing. Profit, the yardstick by which business was measured in the past, is coming under attack. Consumers, environmentalists, insist that the moral standards of business are low, if moral standards exist at all. On the other hand, executives believe, with some justification, that the free market, if not a moral conscience, will keep them honest, that business openly dishonest cannot long exist. The famous "invisible hand" of Adam Smith is still referred to as a modern means of controlling business. Smith, in *The Wealth of Nations,* written in 1776, said that a merchant "when he acts by directing that industry in such a manner as its produce may be of the greatest value, he intends only his own gain, and he is, in this, as in many other cases, led by an invisible hand to promote an end which was not part of his intention."[5] The remainder of the passage is not so well known. Smith continues: "Nor is it always the worse for society that it was not part of it. By pursuing his own interests he frequently promotes that of the society more effectively than when he really intends to promote it. I have never known much good done by those who affected to trade for the public good. It is an affection, indeed, not very common among merchants, and very few words need be employed in dissuading them from it."[6]

What society is now saying is that they "want to be a part of it." It is not sufficient that a company be judged on whether or not it made a profit. Now it is important to know how the profit was made, and value systems are emerging that say that making profits is less important than not making weapons, or not polluting the air and water.

The idea of a social audit is not an easy one for most managements to accept. They want to "do good," but the process of auditing is so foreign to present management activities that the initial reaction is usually that it could do more harm than good. In the first place, the results of the social programs are, in most cases, not measurable in the usual sense. Secondly, the things that are included in social programs are subjective, and capable of different interpretations depending on the value of the observer. Consequently, the business risks criticism if it submits to an audit that might make the company look bad.

However strong the political obstacle might be to agreement on a social audit, the very real practical obstacle is that of measuring results. Aside from the question of whether or not these activities are quantifiable is the question of what standards of performance will be used, and who will set the standards. Examples abound already of admitted social issues that are subject to widely differing evaluations. Pollution is one. Whose values will be used to set criteria for pure air? Those of the company accused of pollution? The environmentalists who want zero pollution? The Environmental Protection Agency who must try to keep everyone happy? The users of the air? The employee whose job might be affected? The academic scientific community?

Pollution is one of the easier issues to analyze. Everyone agrees that industry has the responsibility to keep pollution at a minimum. The minimum hasn't yet been agreed upon. The cost to an individual company can easily be calculated. Benefits cannot be so easily defined in dollars. The consequences of polluting can be defined but, in most cases, cannot be calculated. Therefore, in spite of the measurable aspects of the problem, a true cost/benefit relationship cannot be stated in the way that would be required for a true audit.

Most of the issues that are classed as social are involved in some way with the "quality of life." Since this is another undefinable factor, it does not add much to the solution of the problem of measuring benefits. Each individual responds differently to different situations, whether they be mental or physical. The physical environment is certainly one of the factors that affects an individual's perception of the quality of his or her life. Pure water is a totally different ideal for the affluent vacationer relaxing on a beach or fishing a stream than it is for a production worker whose job might be lost as a result of increased expense accruing to a company that must purify the water.

If a true audit is to be put into effect, some way should be found to record on a company statement the cost of such activities. It is obvious that, if members of minority groups are hired for jobs that could be handled better by non-minorities, the net effect is an increased cost, although it might be difficult to quantify. If a firm converts from coal to a less polluting but more expensive form of energy, some means should be found to credit the firm for that social gesture. If a firm, in planning construction of a new manufacturing unit or office site, includes in its plans, at an increased cost, recreational areas for the use of its neighbors, the increased cost should be credited to the social actions of the company.

I don't believe that it will ever be possible to measure the benefits resulting from social action in the same quantitative way as, for example, the benefits from a sale. But this has been true for many areas of business since the beginning of business. It is not possible to measure the real return for most of what are called staff activities. What is the return from a long-range planning group, or a personnel group, or a public relations group? They don't make a product, nor do they sell a product. It is easy to accept these unmeasurable activities because they are thought to be necessary throughout business. The fact, then, that social programs cannot be measured should not be considered a valid reason for not engaging in programs. It is true that various evaluations and interpretations will be given to the programs. It is true that a real audit cannot be run. However, a simple listing of what a company thinks to be its contributions will at least be a start.

There is no doubt that public concern over social activities of corporations is increasing, and is finding favor not only with young activists but with many of the most sophisticated investors and investment groups. A prospectus issued by the Dreyfus Third Century Fund (May, 1971) specified that it was "seeking capital growth through investment in companies which . . . not only meet traditional investment standards but which also in their corporate activities show leadership in, or have demonstrated their concern for, improving the quality of life in America . . . Activities in the area of the protection and improvement of the environment, and the proper use of our natural resources, and the health, education, and housing demands of America will be considered by the fund in its selections . . . ."

Demand for information concerning business activities has become so widespread that separate companies are now being formed to supply such analyses. One of these, the Council on Economic Priorities, has been in existence for about four years, but already has been a force for change in industry. CEP has four special areas of interest: pollution control, minority employment policies, foreign investments, and involvement in the production of war materials. The Council has supplied information to corporations, universities, government agencies, action groups, and individual investors. The group has issued a report describing 105 corporations that manufacture antipersonnel weapons, and has published a comprehensive review of paper producers, with particular emphasis on pollution cause and control. Prestige is added to the efforts of the group by the support of such consultants as Barry Commoner and Nobel laureate George Wald.

Several years ago, the magazines *Washington Monthly* and *Institutional Investor* sponsored a conference on "Profits in the Public Interest," attended by about 100 investment managers and advisers. One of the questions put to the conference was: "Company A racks up a 20% annual growth in profits, but it won't hire Blacks and its factories pollute the air. Company B is growing only 15% annually, but its reputation on social problems is good. You manage investments for a pension fund. So, which stock do you buy?" The unanimous decision of the panelists was to buy Company B. The rationale of the managers was that Company A was a risky investment, and that it would eventually be forced to change its policies with some sacrifice of profit.

University investment officers and church groups were among the first to proclaim their responsibility for trying to influence the social policies of corporations by speaking out at a stockholders' meeting. Much of their concern was directed at companies making materials for the support of the Vietnam war, and at those international corporations operating in foreign countries where natives were being discriminated against.

The latest attempt to encourage greater social responsibility involved the use of proxy voting to force the consideration of socially-oriented issues as agenda items at annual stockholder meetings. A survey conducted by the Investor Responsibility Research Center, a Washington-based group that researches socially-oriented proxy issues, turned up 90 resolutions in 1974 dealing with social issues, nearly double the number of 1973. In 1976 the number of issues presented was 133, in 1977 it was 77, in 1978 it was 95 and in 1979 it increased again to 126. In addition, they reported that in 1979, 28 proposals had been withdrawn before votes, when companies agreed to reconsider their original positions. The issues involved such things as operations in South Africa, publication of data on Equal Employment Programs, strip mining operations, political campaign contributions and political activities, and female representation on the Board of Directors.[7]

While the shareholders were successful in getting some of their proposals accepted, the mere fact that they were considered for proxy vote is a big change. It means that managements were taking the position of socially minded stockholders seriously. It was not too many years ago that the demand to have a proposition backed by less than 1% of the stock vote considered would have been ignored. Under SEC rules, if a proposition is voted upon and receives less than 3% of the votes, it cannot be reconsidered the following year. If 3% of the votes are cast in favor of the proposal, it can then be reconsidered the following year. At that time, 6% is required for reconsideration, and a vote of 10% required for the following year. In 1974, 15 resolutions listed in the IRRC report received the necessary 3% that now enables the sponsors to keep the issue before the public at the next shareholders meeting. In 1979, 68 out of the total 98 issues voted on in the proxy statements received enough votes to keep the issue alive. The IRRC report also points out the interesting fact that, when the voting records of universities, foundations, banks, and insurance groups were analyzed, at least some in each group supported some shareholder proposals, or abstained from voting to protest a management position. This includes the Ford, Rockefeller, and Carnegie Foundations.

Since the beginning of what we now recognize as industrialized United States, business has done a remarkable job in meeting its basic objective. The railroads, starting about the middle of the last century, allowed business to think of operating outside narrow geographical limits. The introduction of steam for power in factories allowed production on a scale and with an efficiency not possible before. At about the turn of the century, more and more companies were moving toward the integrated organizations we know today—organizations that control the source of the materials used in their manufacturing, that had their own method of distribution, that were reasonably self-contained.

But all business grows or fails as it serves the public to a better or worse degree. All business executives know, whether they articulate the principle or not, that they are in business to serve the public, and it is this service to the public that I submit has been accomplished remarkably well. In general, business has produced quality products at prices the customers could pay. If individual companies did not do this, they failed. Business has provided most of the wealth of the nation, and has made that wealth available through supplying jobs for between 80 and 90 million people, through the payment of taxes that largely support our government, through direct payments of dividends to more than 30 million shareholders, and through indirect payments to another 100 million through pensions, mutual funds, and insurance policies.

The top 500 large U.S. corporations produce about two-thirds of all industrial sales. But the contribution to our economy is not limited to these giants. There are about 9 million individually-owned businesses and farms operating today, an increase of about 1 million in the last 15 years. The number of corporations has doubled from 750,000 to 1.5 million in the same period of time. The voluntary assumption of expenses involved in social enterprises is a drain on company resources that has increased rapidly in the last decade. It has had, obviously, an effect on company earnings opposite to what executives seek.

Anytime a company contributes to a community project or gives any kind of charitable donation it is taking money directly from profits. It has been argued that such contributions are really investments, that the company will, in some way, realize a profit return. I don't believe many such activities ever result in profit. In most cases customers have no idea of what are called "socially responsible" activities of a company supplying a product they are buying. Customer choice is based on price, quality, and, in many cases, effective advertising. The reason companies contribute to their communities, to charity, to the arts, or to education, is that they believe they do have some responsibility to advance the social standards of the country and, a not inconsiderable factor, a company management would appear small in the eyes of its peers in other companies if it did not take part in this generally accepted practice.

The way in which specific contributions are made is an indication of the profit significance companies attach to this activity. A contribution is made to a college because an officer or a member of the board has an interest in the college. The United Fund receives increased support because a company volunteer is head of the drive that year. Contributions are made to a single association of colleges so that an excuse will be available to refuse gifts to the individual colleges.

If charitable contributions were really profitable, why isn't there more charity? The tax system allows the deduction of up to 5 percent of a company's pre-tax income for such purposes. Yet, as mentioned, companies that do have programs still contribute less than 20 percent of the allowable deductable limit.

My own position is that "all of mankind" is not the business of business. Business does influence all of us, but it does so by making it possible for individuals to fulfill their own ambitions, and it makes it possible for nations to establish higher or lower economic goals, depending on the state of advancement of industry in those nations. Companies do recognize that they owe more than daily wages to their employees, and that they owe something to the communities in which they operate. But to believe that industry should be responsible for solving the social ills of the world is unrealistic. Industry can be a major factor in helping solve some of these ills. But it can make its greatest contribution by being successful. Success is not defined either by the description of Marley or Polybius. The successful company will be judged not just on how much profit it makes or how much money it contributes to education and charity. If a company is to contribute its share to being a part of solving social ills, of contributing to an economy that can be a leader in world economies, to an increase in the quality of life of those involved in its operation and in using the results of its production, it will be judged on how creative it can be in doing those things that make it possible to be different.

Being different does not necessarily mean being better. But being the same by definition means that a company is not better. In business competition there is intense interest in what the other company is doing. If you can beat competition to the punch it is a triumph. But there are still many companies, profitable companies, that have a stated policy of never being pioneers, of never being the first to introduce a new product. They simply take advantage of others' agreeability to take risks. It is a safe approach. If the product fails, nothing is lost. If it is a success, the second company jumps in and tries to compete on whatever basis is available. Such a company can never be different, can never make a major advance, can never be creative. It can, however, be successful if success is measured only in profits.

There is little true creative change in industry today, due to both internal and external factors. Machiavelli said: "There is nothing more difficult to carry out, nor more doubtful of success, nor more difficult to handle than to initiate a new order of things. For the reformer has enemies in all those who profit by the old order, and only lukewarm defenders in all those who would profit by the new."

Nowhere is this more true than in an industrial organization. "May change strike you" is an old Chinese curse. There is little danger that the curse will fall upon many business leaders. Everyone knows the incantations, the charms, that neutralize evil effects. They are included as part of Status Quo 101 taught in every business school. They have been effective for all but a few leaders who would still rather wrestle with the demons of risk than sleep with the slaves of conformity.

3 . . . wherein business is regulated, and the costs of the regulation direct business decisions, and there is doubt about the utility of the regulations

Show me a business executive who doesn't think there are too many government controls over business and I'll show you a business executive who has been retired for fifteen years, subscribes to *Village Voice* instead of *The Wall Street Journal,* and believes that Ralph Nader is the greatest stimulus for profit since Franklin D. Roosevelt. To believe that business is not over-controlled is un-American.

On the other hand, to think that business shouldn't be controlled is also un-American. I know of no responsible individual in business who would not agree that most of the government regulatory agencies serve a useful purpose, and are required for the protection of the industry and the public.

Next to the conservative attitude of the business executive, and probably contributing a great deal to that attitude, I believe that stultifying government restrictions are the greatest deterrent to business investment, and consequently to risk taking and further consequently to creativity. No one would quarrel with the idea that there must be a series of checks and balances that allow for progress while, at the same time, protecting the various divisions that make up society from being harmed or from harming each other.

There are few businesspeople who do not believe that today there are too many government checks and not enough business balances. Business is now told whom it may hire and fire, what minimum wages it must pay, how prospective employees can be interviewed, what information it may keep confidential, what the racial backgrounds of its employees must be, what it can say in an advertisement, what the quality of its products must be, whether or not it may grow by expansion and acquisition, whether or not it may obtain and retain a major part of the market of a particular product, how its products are to be manufactured, when its employees may retire, where it may locate its plants, and when it may discontinue operating a plant.

Since 1970 a federal bureaucracy employing 80,000 people has been spawned for the sole purpose of protecting consumers and workers from harm. The Occupational Safety and Health Administration (OSHA), the Environmental Protection Agency (EPA), the National Highway Traffic Safety Administration (NHTSA), all created in 1970, the Consumer Products Safety Commission (CPSC) in 1972, and the Office of Surface Mining Reclamation and Enforcement in 1977, are among those with the greatest authority.

But resources required to fulfill the requirements of the agencies are just the tip of the iceberg compared to the total expenditures drained, not only from industry, but from private citizens. No one really knows the total of our economic and intellectual resources that goes to meet what those intimately involved in the process describe as totally unnecessary and unproductive procedures. Simply doing the paperwork necessary to report to the agencies drains hundreds of millions of dollars from the productive economy.

William E. Simon, former Secretary of the Treasury, in his book *A Time for Truth* offers some statistics showing to what extent the requirements of government regulation are choking the operations of business. He makes the blunt statement that: "No one alive even knows how many federal regulations over business there are. To list all the rulings and regulations established in 1976 alone required 57,072 pages of fine print in the *Federal Register*."[1] Mr. Simon had looked at the report in 1976; the number of pages rose to 87,012 in 1980. He refers to the Commission of Federal Paperwork for statistics showing that government agencies print about ten billion sheets of paper a year to be completed by U.S. business. The government then spends $15 billion to process the paperwork. There are all kinds of estimates as to the total cost to business caused by these government requirements. These very widely, but the one common denominator is that they are all huge. The General Accounting Office, which would not be inclined to exaggerate, in a report released on November 21, 1978, said that the federal agencies themselves estimate that business spent 69 million hours a year filling out 2,125 different kinds of government forms. In addition, business required 613 million hours a year filling out tax forms.

A column by Jack Anderson in July of 1978 placed the cost to private citizens and private businesses for merely filling out federal forms at more than $60 billion a year, including an estimated $21 billion worth of working hours spent on preparing individual income tax returns. His figures indicate that in 1977 businesses prepared 15,000 different types of forms totalling 114 million forms. The Internal Revenue Service alone has 13,200 separate forms. And in November, 1979, the Associate Director of the Office of Management and Budget presented some good news and some bad news to the Senate Government Affairs Subcommittee. The good news was that government-imposed paperwork had been reduced by 15% between 1977 and 1979. The bad news was that business and the American public were spending more than 786 million hours to fill out 4,916 federal forms and reports, and that new regulations will probably add 50 million hours more.[2]

But the reporting costs are only a small part of the expenditure necessary to operate under the controls put on business. Willard Butcher, the President of the Chase Manhattan Bank, said in a speech to the Chicago Executives Club in October of 1978: "As a nation we spend $100 billion each year on government regulations and the largest share of that total—$85 billion—is in compliance costs, the price business and ultimately individuals pay to respond to the mandates of forty-one Federal and countless State and local agencies with regulatory powers."[3]

Economist Murray Weidenbaum, former Undersecretary of the Treasury and now Chairman of the Council of Economic Advisers, agrees with the $100 billion figure in a study prepared when he was Director of the Washington University Center for the study of American business in 1979. He further says that the impact of government regulations on consumer goods and services will cost every man, woman, and child in the United States $500 annually, up from about $300 in 1976. For example, meeting government standards adds $666 to the price of an average automobile, and even this figure does not take into account the cost of plant safety and pollution control or the impact of government regulations on component parts.

Congressman Alan Strangeland of Minnesota points out that the outlay to keep forty-one federal regulatory agencies in existence amounted to $4.8 billion in 1979.[4] This is more than double that of five years ago, and more than six times the 1970 figure. The increase for the support is growing faster than the GNP, the population, and fiscal government spending as a whole. In the past seven years the budget for OSHA has soared to $173 million, an increase of over 1,000 percent. Support budgets such as the $62.5 million needed by the National Institute of Occupational Safety and Health and the $42.5 million budgeted for state matching funds further increased the total. In 1971 OSHA had 970 employees. In 1979 total employees numbered 3,086.

Not surprisingly, the estimates of the huge cost to industry are not accepted by everyone. Ralph Nader says that the estimate is "sheer nonsense." He further says that, "Regulations provide the largest source of new jobs in industry today. Regulation has created perhaps a million jobs, and eliminated no more than 15 to 20,000."[5] I would be inclined to agree with the guess that "regulations provide the largest source of new jobs." The significant question is not whether or not the regulations require one million new people for compliance. The question is whether or not these one million new people contribute more, either in savings as a result of increased safety or in positive production, than the cost of the programs.

Figures can be more firm when we look at industries or individual companies within an industry. The Council on Wage and Price Stability reported in October of 1978 that the steel industry must answer to 27 Federal agencies, 20 of which have been established since 1970. By the end of 1976 the agencies had issued more than 5,300 regulations.

General Motors stated that it cost them $1.3 billion to meet regulations. They point out that this is more than it cost to operate the entire Federal government during its first 75 years of existence.[6]

The U. S. Council on Environmental Quality reported that private outlays for pollution control in the United States in 1976 were $6.5 billion dollars higher than would have been the case in the absence of federal ecological requirements. The McGraw-Hill Department of Economics says that, in the same year, costs to business to meet OSHA requirements were $3.2 billion.

Exxon Company is required to file more than 400 reports each year to 45 different agencies. Standard Oil of Indiana maintains 636 miles of computer tape just to store the data that it must supply to the Department of Energy. Eli Lilly was required to submit 120,000 pages of data to the FDA to support a new drug application for an arthritis treatment. Each year this company spends $15 million on paperwork, an expenditure that adds fifty cents to the price of every Lilly prescription.

In October of 1977 Edwin A. Gee, Senior Vice President of duPont, addressing the Houston Securities Analysts, gave some long-range predictions for duPont expenditures. In the next ten years duPont expects to have about $10 billion available for expansion and modernization programs. Of that, air, water, and noise pollution abatement will require an astonishing $3 billion—about 30%, which is up from the 12% estimate for 1977. Mr. Gee estimated that if that $3 billion could be spent on productive capacity it would build the equivalent of 27 new plants and create 20,000 new jobs. By the end of 1979 duPont had the equivalent of 3600 fulltime people engaged in environmental control. In that year 761 million dollars had been spent on pollution abatement facilities, and operating costs for 1979 alone were $292 million. In its 1979 10-K, the annual report that each company must make to the Securities Exchange Commission, duPont confirmed its earlier estimate of projected costs, saying that installation of plant and equipment through 1978 could cost $2.5 billion (1978 dollars).

Expenditure of such a huge amount is bad enough, but duPont estimates that three quarters of that amount will not result in any environmental improvement. It will be wasted. Simply building facilities to protect the environment is not the end of the story. The facilities must be operated. An estimated $1 billion per year will be necessary for duPont to operate these facilities by 1985. The operating cost alone will be about 8% of each sales dollar.

Allied Chemical has appointed a Vice President for Environmental Affairs, with a budget of $90 million. In 1977 the company hired more than 300 professionals, and spent $73.4 million on environmental control. Goodyear Tire and Rubber reported that it spent more than $46 million in 1977 to comply with regulations imposed by seven agencies. This is more than the company's quarterly net income from two of the four quarters of that year. Dow Chemical Company has broken down its costs from government regulations into three categories they believe to be appropriate, questionable, and excessive. In 1976 their total costs were up 27% over 1975, reaching $186 million, equal to about 5.5¢ on each dollar of sales. Those expenses designated questionable or excessive amounted to $83 million in 1976, up 38% over 1975. Paperwork alone cost the company more than $20 million.

More than 20 Federal laws cover the manufacture of chemicals. The newest of these, and one that will probably have a severe impact on the industry, is the Toxic Substances Control Act (TSCA) of 1976.

The chemical process industry (which includes chemicals, petroleum products, paper, rubber, non-ferrous metals, stone, clay, glass, and textiles) spent a total of $5.266 billion on air, water, and solid waste in 1976. The cost remained essentially the same in 1977, and it is estimated that about $6 billion was spent in 1978.

And if you really want something to boggle the mind consider the conclusions reached by the National Commission of Water Quality, chaired by Nelson Rockefeller, that, to meet only the present water pollution control standards under the present laws business would have to spend $670 billion.

Small business suffers particularly because of these requirements. Between $15 and $20 billion are spent by this sector of business on government paperwork. Businesses with 50 or fewer employees must complete 75 to 80 types of forms.

No industry and no company in an industry is exempt. The degree of control runs the gamut from simple surveillance of small business to what amounts to total control of other industries such as transportation and communications. Originally, the aims were to control the economy of business, and the controlling bodies were usually aimed at specific industries. Land, water, and air carriers were ruled by the Interstate Commerce Commission, the Maritime Commission, and Civil Aeronautics Board respectively. To operate in any of these areas it was necessary to obtain a license from the appropriate body.

Professor Paul W. MacAvoy of Yale has studied the effect of controls and regulations on American industry. He concludes that direct price controls in fields such as transportation and communication now apply to about 10% of the gross national product. He further concludes that the new regulations, particularly in the health and safety field, "cover another 20% of industry so significantly that for all intents and purposes that sector is also regulated."[7] In the past decade, then, what amounts to direct control has extended to about 30% of the American economy.

In mid-1979 H. C. Wainwright & Co., a Boston consulting firm, announced an analysis of the future of the automobile industry in view of the relative competitive burdens brought about by the industry's safety, pollution, and energy regulations. It is easy to conclude that, on a relative basis, small companies have a more difficult time recovering investments required by regulations than do large ones because of the economics possible in large-scale production. However, the Wainwright report produced a startling conclusion that even large companies could be in trouble in a competitive situation. It concluded that the regulatory cost between 1979 and 1985 when both mileage standards and emission standards will take effect may force Chrysler out of the automobile business. It further states that: "American Motors likely will not be able to continue as a producer of domestic automobiles" and that: "Ford's viability as a full-line auto manufacturer also will be impaired." General Motors would remain as the industry. This situation was brought about because the huge research and development costs are the same regardless of the size of the company. It is logical, then, that the largest company would be the one able to recover these costs easiest, thus resulting in what would be a monopoly situation in an industry.

Gary Hart, senator from Colorado, gives a dramatic example of the hurdles one company in his state had to overcome to engage in a vital energy-producing activity. The company, wishing to make oil from oil shale, had to obtain eighteen permits from federal, state, and local agencies. In addition, the same company needed twenty-seven more permits to begin a housing project adjacent to the operation.

In 1968 Exxon paid over $200 million for the right to drill for oil in one section off the coast of California. In 1969 oil was discovered, and Exxon prepared for development. Their plan was approved by the United States Geological Survey. Twelve years and $400 million later there is still no oil being pumped. At a time when we are sending over $40 billion abroad to pay for foreign oil, what is holding up the production of an estimated 500 million barrels of oil in our own country? Exxon lists the following actions that have taken place since 1968: three major governmental impact studies, twenty-one major public hearings, ten major governmental approvals (and many minor ones), fifty-one consultant studies costing more than $2 million, twelve lawsuits, and a county-wide referendum in which the citizens of the county approved Exxon's operation. Meanwhile, we remain a captive of OPEC.

One of the silliest intrusions of the government into established life styles or cultures is the attempt to set quality standards and specifications on adobe brick, material which has been used for construction for nine thousand years in the Middle East, five thousand years in South America, and for centuries in the Southwest United States. The Department of Housing and Urban Development now insists on enforcing construction guidelines that were never meant for adobe brick. If adobe is to be approved for construction, it must be made waterproof with additives, and the walls must be insulated. The final absurdity is that when an adobe house is built to these specifications it becomes so costly that HUD will not finance it.

There are many mechanisms set up within the government to investigate the safety of a product or process, and to approve its use. The Food and Drug Administration, OSHA, and the U.S. Department of Agriculture are examples. It is unfortunate that consideration of safety and hazard are no longer based only on scientific data. Now, all regulatory decisions are made with one eye on the political implications of a decision, the other on the scientific aspects. When there is disagreement, politics wins. Business, producer of the product, and the general public, the consumer of the product, have been set up as adversaries. Business is perceived as having no qualms about foisting off on the public inferior, overpriced, even dangerous products. The public, therefore, must be protected. The pervasive attitude seems to be that it is safer to regulate than not to regulate, even when, by its own admission, an agency does not know the hazards involved in the regulated action. As an example, the EPA proposed to regulate oil production brines, drilling muds, and crude oil residues as "hazardous waste." The industry viewed this as having a devastating effect on its operations, estimating that it would cost $45 billion in the first year plus an annual cost of $10 billion or more in succeeding years to meet EPA's specifications. The EPA estimated the cost to be as little as $100 million a year.

It added little to the confidence in EPA to read from EPA itself, as part of the proposed regulations the following: "The agency has very little information on the composition, characteristics, and the degree of hazard posed by these wastes." Yet, instead of obtaining information on whether or not a hazard existed, regulations that could severely hamper the search for oil and gas were proposed. In defense of its position EPA sent to the House and Senate committees its basis for its position. It said that all its information on hazards had been obtained from telephone conversations held over only three days, April 23-25, 1979, four months after the regulations had been proposed. It qualified its conclusions by saying: "Much of the information is based on hearsay and thus cannot be validated either by the agency or the agency's source at the time." The House Commerce Committee stripped the EPA of its regulatory authority over oil and gas wastes. The Senate Committee on Environment and Public Works prohibited the agency from issuing any regulations in the area for two years.

Obviously it is a popular position to take, that of protecting untold millions of people from danger resulting from action of their neighbors, of industry, or even of themselves. It should be no surprise, then, that politicians, investigative reporters, and consumer advocates can define hazards to the public that the public doesn't realize are hazards. There is no disagreement with the absolute need for regulatory agencies. But there can be considerable disagreement with the way the lives of individuals are being regulated by these agencies. If it is desirable to protect, then the more protection a bureaucracy gives the better it is doing its job. Such seems to be the prevalent opinion, at least in the bureaucracies.

While no business escapes the net of some government control, those companies based on technology, in particular those based on the physical or biological sciences, face special problems from regulatory agencies because both the products produced and the processes used to produce them have a special potential for danger. Both those engaged in production and the ultimate consumer must, therefore, be considered as potential victims exposed to these hazards.

The sweetening characteristics of aspartame, a chemical resulting from the combinations of two naturally occurring amino acids, were discovered in 1965. After six years of research, the company that had made the discovery applied to FDA for approval to market the product. After a year and a half of evaluating the data FDA did approve the product for sale. However, a short time after the FDA action, a professor of psychiatry and neuropathology at Washington University in St. Louis, and a Washington consumer attorney objected to the approval because they said that animal tests showed the product might cause all sorts of neurological effects, including mental retardation. The FDA then stayed its approval and asked Universities Associated for Research and Education to validate the company's studies. The studies were validated. In August of 1979 the FDA established a three-member body of inquiry composed of scientists to further evaluate the data. A year later, fifteen years after the discovery of the product, it was still not on the market in the United States, although it had been made available in France, Belgium, and Luxembourg. Canada had recommended approval, and nine other countries were considering approval.

During the time that FDA was considering approval, construction of a plant to manufacture the product had been started. By the time the FDA had stayed the approval previously given, the company had invested about twenty million dollars in the plant. This does not include other large sums spent for research, development, and testing. While the company awaits some action, it is spending about ten thousand dollars per day in expenses, not to mention the loss of profits as the sweetener remains off the market.

With these kinds of risks to be faced, is it any wonder that companies are reluctant to invest in very expensive projects whose future success might be determined, not by logic and scientific judgment, but by emotion and fears of public criticism?

Consider what happens when a new drug is being considered for approval. The FDA evaluates all the medical and scientific data available. It then must decide whether or not the data support the claims made for safety and efficacy. If they do, the drug is approved for use. If they do not, it is rejected, usually with the request for more supportive data. During the clinical trial of the drug, it might be tested on 10,000 or 50,000 or 100,000 people with few or no side effects. When it comes into general use millions of people are exposed to it. It is not unusual for side effects of a drug, not detected in clinical trial, to show up during later use. Sometimes these effects are severe enough for the FDA to require additional warnings, or even to withdraw the drug from the market. When this happens, the FDA is subjected to severe criticism for having approved a dangerous drug in the first place, even though it would not have been possible to detect the side effects in the clinical trial.

In 1958, an amendment sponsored by Congressman Delaney from New York was added to the Food, Drug, and Cosmetic Act. The amendment, known generally now as the Delaney amendment, specified that no product could be added to food if the product could be shown to cause cancer in humans or animals. There are no qualifications. It made no difference that the product produced its cancerous result only if administered in millions of times the dose to which a human being would be exposed. It made no difference that the animal test had to be run using injected material when the actual use would be by oral ingestion. The ban was uncompromising and all inclusive. The fact that the amendment relates only to cancer exaggerates the problem, because cancer is such an emotional subject. Not only is it emotional, but because the disease itself is so little understood it is easy to build up a fear of anything even remotely suggesting its cause.

It is natural that anything that can be done to prevent cancer should be done. The mere thought of being exposed to anything that might cause this dread disease strikes fear in the hearts of the general public. Cancer is looked upon as a mysterious disease. Its symptoms may appear fifteen or twenty or twenty-five years after exposure to the causative agent. Its biological cause is unknown. And it has been stated by some that the tiniest part of a carcinogen, even a single atom, may cause the disease. It is no wonder, then, that the most severe environmental protection restrictions have been reserved for things suspected of causing cancer. And it is no wonder that the easiest way to have a product restricted is to say that "it might cause cancer." The statement that "80% to 90% of all cancers are environmentally induced" has been repeated so often in the last few years that it is now generally held to be true. In the way that the statement was originally presented it is true. Unfortunately, it has been the subject of much misinterpretation. The statement seems to have originated in a report of the World Health Organization Expert Committee on the Prevention of Cancer and was published in 1964. It said that environmental or extrinsic factors directly account for more than three-quarters of human cancers and therefore "that the majority of human cancer is preventable."[8] From there the statement has evolved to "some 90% of cancer in people is due to chemicals," or "there is now growing recognition that the majority of human cancers are due to chemical carcinogens in the environment."

The difficulty has arisen from the interpretation or misinterpretation of the word "environment." The word has come to mean, in our vocabulary, water, air, and, in some cases, food. Thus, when we speak about something being caused by the environment, the picture of a pollutant, usually a chemical, comes to mind. However, the correct meaning and the meaning intended by the World Health Organization includes all those things outside our bodies. These things include even cultural habits. Diet, smoking, and exposure to sunshine certainly contribute to the cause of cancer, as indicated by the different rates of cancer in different geographical areas.

Dr. John Higginson, the Founding Director of the World Health Organization's International Agency for Research on Cancer (IARC), was the formulator of this idea. In September of 1979 he gave an interview to *Science* in which he explained how his ideas had become distorted. His first work was done in the 1950s, when he compared the incidence of certain types of tumors in Africa and the United States, and it was as a result of these studies that he first concluded that about two-thirds of all cancers had an environmental cause and were therefore theoretically preventable. As Dr. Higginson says: "When I used the term environment in those days I was considering the *total* environment, cultural as well as chemical. By cultural I mean mode of life. When we ran the study of Blacks in 1952, we started looking at their diets, how they lived, the number of children they had, the age of menopause, the age of menarche—all that was included in the term environment."[9] He argues forcefully against the adoption of too simplistic a view of the cancer process.

By whatever standards one uses, the conclusion must be reached that there is no "epidemic of cancer" in this country. For the last fifty years the age-adjusted incidence for all cancers has remained nearly constant. Only in respiratory tract cancers, stomach cancer, and cancer of the uterus has there been any dramatic change. Respiratory tract cancer, as is well known, has increased greatly, probably as a result of smoking. However, there has been an equally dramatic *decrease* in stomach cancer and cancer of the uterus for as yet unexplained reasons. If, then, smoking was removed as a cancer cause, the incidence of overall cancer would have decreased. This in spite of the increasing age of the citizens of the United States, and the decreasing mortality from other diseases of the elderly. This judgment is concurred in by John Cairns, the Director of the Imperial Cancer Research Fund's Mill Hill Laboratory in London. In his book, *Cancer, Science, and Society,* he writes: "Despite a lot of publicity there is little evidence that the chemical industry causes much of the current total cancer incidence. . . . with the exception of lung cancer none of the common cancers are much commoner than they were fifty years ago, whereas most of the chemicals people are worried about were introduced only after World War II."[10]

The male residents of the state of New Jersey have a

higher average death rate from cancer than do males in other parts of the United States. Because of the high concentration of industry, particularly the chemical industry, in that state, chemicals were again suspected as the cause of the cancers. An epidemiological study of one of the larger chemical producers, Union Carbide, located in Bound Brook, New Jersey, was conducted by Robert W. Mack and Lynne B. Harrison of the University of Pennsylvania Medical School, under a grant from the National Cancer Institute. Not only was the incidence of cancer among these workers lower than the New Jersey average, but it was also below the rate expected for the comparable U.S. population.[11]

An even more definitive study was conducted by Judith A. Zack, of Monsanto, and Raymond R. Suskind, Director of the Institute of Environmental Health at the University of Cincinnati Medical Center. Dioxin, or TCDD, has been suspect as a cancer-producing agent for some time. Since 1940 workers have been producing Dioxin in a plant at Nitro, West Virginia. They were not only exposed to what normal working conditions would be, but also survived an in-plant explosion in 1949, when large amounts of the chemical contaminated the plant. An epidemiological study showed that workers at the plant, followed for a period of more than thirty years, had a lower than expected rate for the United States population either from cancer or from cardiovascular disease.[12]

The International Agency for Research on Cancer estimates that occupational chemicals could have caused from 1 to 5% of the cancer deaths occurring in 1978. Obviously what can be done to reduce or eliminate these deaths should be done. At the same time it should be recognized that conditions as they exist now do not cause 1 to 5% of the deaths. Those people who died in 1978 probably were exposed to cancer-causing environments fifteen or twenty or even twenty-five years before. Conditions have changed drastically in that time. Therefore, some rational evaluation of the relationship between environmental factors causing cancer now and those existing years ago should be made. Otherwise, we shall find ourselves shooting at where the target once was, spending millions of dollars and inhibiting progress while, at the same time, not having any effect on true prevention of cancer.

A striking instance of regulatory action affecting the general public came with the banning of the synthetic sweetener cyclamate in 1969. Cyclamate is a non-caloric substitute for sugar. It had been used for over ten years before 1969 by diabetics in diet drinks and foods and as an alternative for those who simply did not want to eat sugar. Its use was unrestricted, and there were no observed harmful effects on human beings. Then studies on rats, using massive doses of cyclamate daily for the entire lifespan of the rats, turned up an "unusual" form of bladder tumor. The amounts of cyclamate used would be equivalent to an adult drinking 500 10-ounce cans of soft drink every day from birth until death, or of a 40-pound-child drinking about 140 cans per day. Ten years later, in spite of presentations of massive data by the manufacturer of the product, cyclamate is still not approved for use. It now appears that the cyclamates will never appear again on the U.S. market. In September of 1980 the FDA, after evaluating data collected since its original ban in 1970, again refused to approve the product for use. At this point the only company in the United States producing cyclamate decided to give up its effort to re-introduce the product because, as it said, it was impossible to meet the scientific proof demanded. Thus, a product deemed to be safe by Norway, Denmark, West Germany, Switzerland, and other European countries will not be available in the United States. In Canada, cyclamate had been banned in 1969. On October 5, 1978, the Ministry of Health and Welfare reversed that decision and said that, since cyclamate is not carcinogenic, it could be used to sweeten drugs.

In 1976 the Delaney amendment struck again. A group of Canadian researchers reported that massive doses of saccharin cause cancer in rats. Immediately, the FDA proposed removing saccharin from the market, stating that they had no alternative in view of the Delaney amendment. This time there was strong public reaction. In most cases the public is never aware of what it is being deprived of or protected from. When cyclamate was withdrawn there was some protest, but the cyclamate users quickly turned to saccharin as an acceptable sweetening agent. Now, if saccharin were eliminated, there would be no non-caloric sweetener available. This would have an immediate effect on millions of people accustomed to diet drinks and non-sugar-sweetened foods. The protests were immediate, this time not from scientists but from the consumers.

It was stated that removing the last sweetener from the market would do irreparable harm to diabetics and weight watchers. For others, it would mean a return to sugar, a product that in itself has been shown to have serious side effects. With public pressure mounting, Congress got into the act and a bill was passed preventing the FDA from removing saccharin from the market for a period of eighteen months, in spite of the Delaney amendment. When the first eighteen month moratorium had passed without any action, Congress voted an additional two year extension, until June 30, 1981. The vote in the House was 394 to 22 in favor of saccharin. The Senate passed the measure by a voice vote.

Canada had already banned saccharin. Thus we have the situation where in Canada a patient could get drugs flavored with cyclamate but not saccharin, while in the United States the reverse is true.

After the Congressional decision, the Food and Drug Administration, through its National Center for Toxicological Research, issued a 700-page report that concludes that the possible benefits of saccharin may outweigh its potential hazards. For a government document the report contains some startling statements. For example, it has stated that the dose of sugar equivalent to a dose of saccharin in a 12-ounce diet soda carries 375 times the "risk of cancer in animals." Also "five salted peanuts would about equal the risk of a can of diet drink." Further, "A typical steak has about twenty times the bladder cancer potentiation of a can of diet soda" estimated from rat data. The report does state that "the greatest potential risk from saccharin probably lies in its ability to promote several different bladder carcinogens." In other words, saccharin might not be a cancer-producing product in itself, but does stimulate other products to become more active as carcinogens.

In early 1980 two scientific articles appeared that indicated that there was no risk of cancer to those ingesting artificial sweeteners. Researchers at Harvard, publishing in the *New England Journal of Medicine,* said that: "as a group, users of artificial sweeteners have little or no excess risk of cancer to the lower urinary tract."[13] Workers at the American Health Foundation published their results in *Science,* and concluded: "No association was found between use of artificial sweeteners or diet beverages and bladder cancer. No evidence was found to suggest that artificial sweeteners or diet beverages promote the tumorigenic effect of tobacco smoking."[14]

As a sidelight to this controversy the history of saccharin is interesting. It has been in use for about eighty years, but it was one of the first two products, sodium carbonate being the other, proposed to be banned by the newly passed Pure Food and Drug

Act of 1906. However, the objection at that time was not to its cancer-causing properties, but to the fact that it was being used generally in food processing in place of sugar and was, therefore, seen as causing a nutritional problem.

The emotional atmosphere in which the decisions were made about the drug Laetrile is another example of a clash between politics and science, resulting in inconsistencies of regulation. Laetrile, an extract of apricot pits, has received a tremendous amount of publicity as a cancer cure. It has been available in Mexico, but has been banned in the United States. As a result, thousands of cancer sufferers have journeyed to Mexico to be treated. Additional thousands have used smuggled material for their treatments here. The drug cannot be approved by the FDA because there is no scientific evidence to show that it is safe and effective. All the tests usually run in determining effectiveness have been negative. However, because of the claims from some patients that they had been cured by Laetrile, a massive public opinion pressure has built up to make the drug available. Twenty-one state legislatures have approved the drug for manufacture and sale in particular states, and legislation is being considered in several others.

In order to placate the supporters of Laetrile the National Cancer Institute surveyed the country to try to find patients who had been treated with Laetrile with claimed benefits. It has been claimed that about 70,000 American cancer patients have taken the drug. In the search for these patients announcements appeared in medical journals and daily newspapers. The National Cancer Institute mailed over 450,000 letters to pro-Laetrile groups, physicians, and health professionals, asking them to collect the case histories. Possibly because of the antagonism of the Laetrile proponents who boycotted the program, only 93 patients were found, and of these only 22 met the standards set up by the Institute, two of which were that cancer had been definitely diagnosed, and that only Laetrile had been given as a treatment. After intense review of these cases the National Cancer Institute reported that "no definite conclusion could be drawn." Six of the patients had shown an improvement, nine had remained the same, and six had gotten worse. The institute decided that "it is impossible to tell whether Laetrile was responsible for the improvement." However, in spite of this, the National Cancer Institute's Decision Network Committee, by vote of 14 to 11, recommended to the National Cancer Institute that a clinical trial on cancer patients be instituted. Two days after the recommendation was made it was accepted by the Institute.

Following a toxicity trial in rabbits, the FDA in early 1980 approved a trial to be carried out in four different cancer centers around the country on a total of 200 or 300 patients. The patients will be volunteers who have not been helped by any other presently known method of treatment. It is assumed that it would require about a year after the tests began in the spring of 1980 before any conclusions can be drawn regarding efficacy.

Consider what would have happened had an ethical drug company tried to sell a product, particularly an anti-cancer drug, on the basis, not of scientific evidence, but of testimonial evidence from patients. It would have been pilloried by the very bodies that are now reacting to public pressure—and with justification.

The Laetrile situation became such a public interest issue that the Supreme Court accepted a case to decide whether or not the personal freedom of an individual had been abridged. A terminally ill cancer patient in Oklahoma claimed that the FDA's ban on the interstate shipment of the drug had prevented him from having available a possible cure for his disease. He won victories in the local court and later in a Federal Appeals Court in Denver, where the judge found the FDA had virtually no authority to control drugs sought by terminally ill cancer patients. The rather tortuous reasoning employed by the court revolved around the fact that the FDA by law must base its decisions on whether or not a drug is safe and effective. By definition a terminally ill patient is one for whom there is no effective drug. The court ruled: "Therefore, we hold as a matter of law that the 'safety' and 'effectiveness' requirements of the statute as now written have no application to terminally ill cancer patients who desire to take the drug." Presumably the same reasoning could be applied to any terminally ill patient who would then be free to take any drug, or expose himself or herself to any treatment, legal or not.

In June of 1979 the Supreme Court announced its conclusions. In an unanimous decision the federal government's ban on Laetrile was upheld. The court said: "The Federal Food, Drug, and Cosmetic Act makes no special provision for drugs used to treat terminally ill patients."

The issues involved in just these two cases—saccharin and Laetrile—involve a complex interrelationship of scientific, political, social, and even moral issues. To solve the problem on one basis is likely to violate the principles based on others. In both cases it would appear that Congress was exercising scientific judgment.

Should Congress be able to overthrow a scientific decision of the FDA? I believe the direct answer to that should be no. Having said, as a matter of policy, that carcinogens should be excluded from food, I do not believe that Congress should then be put into the position of deciding what is or is not a carcinogen. This is strictly a scientific decision.

Should Congress be able to decide that saccharin should stay on the market? I believe the direct answer to that is yes.

My answers to these two questions seem contradictory. But let me explain. When Congress passes broad legislation it sets up principles under which the appropriate regulatory agency operates. It may include classification of products to come under control, as it did in the Pure Food Law and the Toxic Substances Act. It may be specific or nonspecific about limits. However, once the law is in effect, it is then the responsibility of the agency—FDA, USDA, OSHA—to administer the law. The administration involves making scientific decisions required to decide whether or not a product or process fulfills the requirements of the law. Is an animal test appropriate? Are the statistical analysis results significant? Are there satisfactory measurements for determining the concentration of a product in the air? Such decisions are strictly scientific.

In the case of saccharin, then, we seem to have Congress exercising scientific judgment. Congress had passed the Delaney amendment in the first place. FDA had decided that saccharin was a carcinogen. Now, we have Congress being in doubt as to whether or not it is a carcinogen. Deciding on carcinogicity should not be part of the democratic process. You don't vote on whether or not something is dangerous. It is another matter if, recognizing danger, you vote to make the conscious decision to risk the danger. In this case I think the result of the Congressional action was correct. In any practical amount, I don't think there is any danger from saccharin. But I think it is a dangerous precedent for Congress to express scientific judgment, to try to rectify the mistakes brought about by their own stupid law. Incidentally, the vote in Congress in favor of allowing saccharin to stay on the market was 375 to 23. The sensible approach for

Congress to take would be to admit that there are carcinogens too weak to be dangerous, or that there are levels of carcinogens that are non-toxic.

The decision regarding Laetrile involves a completely different analysis to approve its use. It is necessary to say that scientific procedures now used to justify even the initial clinical testing are useless, and that some other basis for predicting activity must be used. None of the pre-clinical tests required for normal drug approval has shown positive results when applied to Laetrile.

However, even after the data have been evaluated from a scientific point of view a major decision remains. What is the definition of "safe" and what does "effectiveness" mean? It seems to me that, in overriding the decision of FDA on saccharin, Congress was, in a way, giving their own definition of safety. In fact, in reversing the FDA decision they arrived at the Solomon-like process, not necessarily to contradict FDA, but to postpone a decision on the matter for 18 months, during which period of time additional safety tests could be run.

In many cases the FDA has been blamed for actions over which they had no control. Once a law is passed by Congress, FDA must administer it whether it is good or bad.

No less an authority than Dr. Donald Kennedy, the former Commissioner of the Food and Drug Administration, has compared the food additive provisions of the Food, Drug, and Cosmetic Act to the Winchester Mystery House in San Jose, California. The builder of the house was told by a spiritualist that as long as she kept a building project going she would not die. After 38 years of building, she had a house with one hundred and sixty rooms staggering over six acres with doors leading nowhere, blind chimneys, towers, closets opening into space, stairways with no upstairs. The owner died in the midst of a further project. The food additive provisions have had a similar history. Over the years since 1958 the laws and agencies interpreting the laws just grew. Congress got into the act by passing laws, some of them totally non-scientific, some conflicting with already existing laws. It was, then, up to the FDA or the USDA or the EPA to administer them. Thus the mystery house grew.

Possibly as a result of the public reaction and Congressional action involving saccharin, the FDA in at least two cases seems to be taking a more analytical approach to the control of carcinogens. It has been know for some time that a class of chemicals called nitrosamines causes cancer in animals. It has also been known that nitrites, under certain conditions, are converted into nitrosamines. Nitrites have been added to meat as preservatives for years. However, it was believed that the nitrites themselves were harmless. In August of 1978 a study from MIT strongly suggested that "nitrites produced cancer of the lymphatic system of test animals. The mechanism is clearly distinct from that of nitrosamines."[15] This should have been enough to result in an immediate ban on the use of nitrites. The situation is complicated. Processed meats, fowl, and fish containing nitrites account for about 7% of the American food supply. However, only 20% of the consumption of nitrite comes from this source, the remaining 80% coming from beets, radishes, spinach, leafy vegetables, celery, and drinking water by way of conversion from nitrites to nitrates. Average levels of thirteen parts per million in dry beans to 2,760 parts per million in beets have been reported, while some radishes have been found to contain as high as 15,000 parts per million. Cured meats contain from 10 to 40 parts per million of nitrites.

The body itself produces in saliva, the stomach, and the intestine orders of magnitude more nitrites than are ingested from cured meats. Doctor Steven R. Tannenbaum, Professor of Food Chemistry at MIT, says: "About ten times more nitrite enters the stomach daily from the reduction of nitrate in saliva than from nitrite in cured meats; probably thousands of times more nitrite is formed in the intestine than is contributed to the intestine from pre-formed nitrite in the diet. The risk that arises from the use of nitrites according to present and proposed USDA regulations may be minuscule compared to that resulting from the body's natural processes."[16]

Added nitrite does serve an important function in preventing botulism. The preservative allows cured products to be shipped for long distances without refrigeration. The American Meat Institute estimates that a nitrite ban "would all but destroy the U.S. hog industry since nearly 70% of our pork ends up in processed meat products—mostly cured meats." Saccharin is involved in a $1 billion industry, nitrites in a more than $12 billion industry. So the FDA has moved cautiously, stating that it is "presented with a difficult balance of risk."

In August of 1978 both the FDA and USDA decided that nitrites should be phased out as a food additive, and scheduled a press conference to announce the decision. But a funny thing happened on the way to the conference. The Secretary of HEW, possibly remembering the saccharin situation, decided to ask the Justice Department to review the FDA-USDA decision. The press conference was cancelled. And the Justice Department did, indeed, rule that a phase-in plan for a ban on nitrites was illegal. The alternatives were, then, to ban nitrites completely immediately, or to allow them to be used without limitation.

Congress again entered the fray. With the agreement of the Administration a proposal was written that would delay for at least a year any action on the cancer-causing additive, thus overriding the Delaney amendment. At the same time permission was requested to phase out uses for nitrites over a period of several years as substitutes became available.

At the same time a group of independent pathologists, members of Universities Associated for Research and Education in Pathology, were asked to evaluate all the data on which the suspicion of carcinogenicity was based. This independent review was evaluated by the FDA and an Inter-Agency Working Group on Nitrite Research. In August of 1980, two years after the original suggestion was made to phase out nitrites, the FDA and USDA concluded that there was not enough evidence "to support a conclusion that nitrite induced cancer in rats." This situation will continue to be monitored by the regulatory agencies. But for the present, at least, rational analysis of scientific data seems to have triumphed.

The complications surrounding attempts to meet regulatory requirements sometimes become absurd. Following the publicity given to the potential danger of nitrites, suggestions were made, reasonably enough, to remove them from the market. However, the federal court in Iowa has issued a permanent injunction against the sale of nitrite-free meat under such traditional names as wieners, frankfurters, bacon, or bologna, because such labels "could endanger the public." The National Pork Producers Council had sought the injunction contending that products sold and consumed under the new regulation could cause severe illness and hurt pork's image as safe.

A second case indicating restraint on the part of the regulatory agencies is illustrated by the treatment accorded the presence of aflotoxin in food. Aflotoxin is a poison that is produced when mold grows on certain crops. It has been shown to be a carcinogen. Earlier I mentioned that a government report has suggested that "five peanuts might be more carcinogenic than a saccharin-containing diet drink." Nothing is sacred. Peanut butter is now suspected of being carcinogenic. Aflotoxin is present at some level in practically all peanuts and peanut butter. The FDA has set maximum levels allowable in peanut butter, contrary to the situation occurring with other carcinogens. FDA can do this because the aflotoxin is naturally occurring. By some mysterious logic only synthetic products added to food are deemed to be dangerous enough to warrant complete exclusion.

When Congress in 1977 ordered a hold on the banning of saccharin, at the same time it asked that the National Academy of Sciences review the entire additive problem. In March of 1979 the Academy released its report.[17] It recommended a sweeping overhaul of the Food and Drug Administration procedure for assuring food safety, including repeal of the Delaney Amendment. The report said the present procedures are "complicated, inflexible, and inconsistent." Basically, the Academy suggested a system in which health risks would be weighed against health benefits before a substance could be banned or restricted. Certain effects such as those that might result in cancer or in genetic change even though they might have only a small chance of occurring would have more weight in determining approval than would a mild allergic reaction that would occur more frequently. Whether or not a suitable substitute is available would also influence the decision to ban a product that had some health benefits.

One of the most controversial recommendations was that permitting the FDA to consider economic costs in their decisions. The panel believes that economic factors might be considered on marginal issues. Different members expressed the concept in different words: "Primarily health benefits be weighed against health risk, but that other benefits might be taken into account" or "The risks should be the determinant, with benefits as a modifier" or "The benefits should be more modifying when the risks are low or moderate."

Most regulatory agencies assume that any cost is justified in eliminating any suspicion of risk. They are usually righteously horrified if the suggestion is made that economic factors should be included in considering how regulations are to be implemented. In response to a suggestion that economic analysis of proposed rules be prepared, the OSHA administrator, in November 1978, responded that such analyses "are not what is required in the Occupational Safety and Health Act. I have the law to uphold, that is my mandate." The head of the Environmental Protection Agency in March 1979 said that his agency did not have a profit motive, that he had no stockholders, that he had no shareholders, and no year-end bonus.

It is just such attitudes that contribute to unreasonable regulations. If the objective is to protect the public at any cost, then the simple decision is to remove all traces of all foreign substances, not that have been proven to cause ill effects, but that have not been proven to not cause the effects. I suggest the regulatory agencies do have stockholders to whom they should be responsible, and the agencies do have shares in a company, and they do have a profit motive. The responsibility of a regulatory agency is not just to protect the public from harm. This is the negative aspect of regulation. The equally great responsibility, and one that is all too frequently ignored, is the positive duty to encourage progress, not just to inhibit action. For example, the FDA has a duty not just to keep dangerous or ineffective drugs off the market, but also to encourage the development of safe and effective drugs. OSHA is responsible not just for eliminating dangerous practices, but for encouraging the development of new and safer practices. But none of this should be done without at least recognizing the economic factor that exists in every decision made by every government agency. The safest drug company would be a company that was not allowed to produce drugs. Pesticides could not harm anyone or anything if restrictions became so severe that no chemical could be produced. But the benefits are also eliminated.

As the cost of regulation, not just in money but in lost products, lost jobs, and even lost businesses becomes generally known, a consideration of the consequences is becoming accepted and a justification for the promulgation of regulations is being required. In 1978 OSHA published standards reducing from ten parts per million to one part per million the amount of benzene that could be present in the atmosphere of a workplace. Industry, which had agreed to the existing ten parts per million standard, immediately appealed. The original standard had been established because it was believed that benzene could cause leukemia above this level. It was estimated that the new standard would eliminate one new case of leukemia every six years. In October of 1978 the Fifth Circuit Court of Appeals in New Orleans struck down the standard, saying that OSHA must have "some factual basis for an estimate of expected benefit before it can determine that a one half billion dollar standard is reasonably necessary." OSHA appealed this decision, and in July of 1980 the Supreme Court announced its decision, upholding the lower court. It said, among other things, that safe did not mean risk free. A work place can hardly be considered unsafe unless it threatens the worker with a significant risk of harm. The court also concluded that OSHA had not demonstrated that its original limit of ten parts per million of benzene was unsafe. Therefore, it made no sense to alter the standard by 90 percent at a cost of five hundred million dollars as estimated by OSHA.

Predictably consumer groups suspect a plot by private interest groups to allow for manipulation of government agencies. Ralph Nader's Health Research Group says that: "Corporate wealth would once again take precedence over public health" if the FDA is given the right to make judgments. Dr. Donald Kennedy, who was then commissioner of FDA, opposed the proposal absolutely. He said: "I do not want the power to weigh economic benefits against health risks unless Congress explicitly tells me the value of a life." Dr. Kennedy has indicated that he believes FDA should have some more discretion than it now has. But he also believes that such a discretionary policy would encourage more attempts by lobbying groups to influence FDA decisions.

I do not have much sympathy for a scientific agency that does not want the freedom to make what are essentially scientific decisions. Certainly it is easier to administer a dogma, since no thought, no selection of alternatives, is required. It is the duty of the FDA to make the decisions. Obviously, lives should not be sacrificed for the economic gain of business. But the decisions more often than not do not involve life and death decisions. What economic cost might be required to prevent a hundred or a thousand temporary and even trivial allergic reactions in the population? Or, to put the question another way, is it justified to expose small numbers of people to possible minor discomforts in order to save thousands and even millions of dollars while at the same time providing other health benefits for many? Most decisions do not involve factors either as radical as death or as trivial as minor allergies. But they are decisions that should be made by those trained to make them, and they should be made based on the data available. If Congress is expected to mandate decisions, it can do no more than it did when it passed the Delaney Amendment, namely, be so conservative that it eliminates the possibility of judgment.

The objective of the regulatory agencies is to protect the citizen from risk, whether it be economic risk or physical or mental risk. In some cases such as the Federal Communications Commission and the Federal Trade Commission the agency has even assumed the responsibility for deciding on the moral and social risks. I would be the first to agree that most of these agencies are essential to the operation of our government. In a very objective moment I would even agree that there should be an OSHA or something similar.

Protection may be provided by preventing something from taking place, as in keeping a drug off the market, or it may result from forcing an action to take place, as in the compulsory installation of airbags in automobiles. In both of these types of responsibilities there is grave danger of the regulatory body over-reacting, not just to protect the public from danger, but to protect the bureaucracy from criticism.

Most of the controls that influence business decisions never become subjects for public discussion. Society in general does not know how their lives are being influenced by actions or, in some cases, lack of actions by regulatory bodies. They do not know of the drugs that are not approved or the manufacturing process that must be changed, or the mountains of paper that must be submitted. It is only when an action directly affects the general public, and where the public is either caused inconvenience, or where an individual cost is immediately identifiable, that the citizens of the country have a chance to evaluate how effective regulatory agencies are.

There is no way the FDA can protect everyone against the rare chance side effect of a drug without requiring the testing of the drug on everyone who would eventually be using it. How, then, could the FDA protect itself against future criticism? The first way is to avoid the approval of any drug. It is easy to detect the harm caused by a drug that has been approved. It is possible to assess the harm that might be caused by not having a drug available to treat a specific disease. However, the difference between the two is that, when a drug is approved and causes side effects, it immediately becomes a subject for public discussion, even public notoriety. When a drug is rejected before approval, no one but the sponsor of the drug knows or usually cares about it. If there are questions raised, it is impossible to argue against the FDA position that it has not been approved because it is either dangerous or not effective or both. Thus, we see the situation where, because of a desire to insure total safety, requirements for drug testing have reached the point where it now takes 8–10 years from the time of starting research, and multi-millions of dollars to get a drug to the market.

The result is predictable. The tendency of the regulatory agency is to inaction, rather than action, because it is safer. Those who have, in the past, been responsible for the conception and development of miraculous disease-curing products now find they can no longer afford to invest in research, both because of the cost and because of the uncertainty of having the product approved even after the investment of time and money has been made.

There are many glaring examples of the condemnation of chemicals because of real or imagined dangerous potential without a consideration of benefits from the use of these products. Over the years more lives have been saved and more suffering prevented from the use of penicillin, DDT, 2,4-D, and chlorine than all other chemicals combined.

These chemicals have been responsible for saving millions of lives by preventing diseases, and preventing untold suffering from malnutrition. On the record, penicillin has probably been responsible for causing more deaths than the total of the other three. Yet DDT is totally banned from the market on the basis of the destruction of some useful birds and insects due to improper use. After more than thirty-five years no real harmful effect has been proven in humans when the product is used properly, even though most of the people in the world do have DDT stored in their bodies and have had for decades. It is ironic that in Canada DDT is allowed to be used in only one crop: tobacco, the plant that is documented to kill more people than any other single crop.

As if the almost total control of industry were not enough, the proliferation of different agencies, all acting independently, has exacerbated the problem. Each regulatory group has its own rules and standards that it administers as if it were the only regulator in existence. Sometimes one has the impression of an Alice in Wonderland atmosphere when two agencies attack the same problem from different directions. In one of the meat packing plants, the Federal Meat Inspection Service required that a hole be cut in a meat conveyor line so that inspectors could remove samples of meat from the line for testing. When the OSHA inspector appeared, he immediately ordered the hole covered because it was a safety hazard.

Park's Sausage Company says that the Agriculture Department required its kitchen floors to be washed repeatedly for sanitary purposes, yet the OSHA ruled that its floors must be dry at all times. The construction industry was ordered by OSHA to install beepers on all moving vehicles at building sites so that workers would hear the warning and avoid accidents. EPA then ordered the workers to wear earmuffs to protect them from the noise.

Do the stringent new laws really protect us any more than the old laws? Do they help or hurt? There is an economic definition of overregulation that has nothing to do with how much trouble regulation causes. Overregulation is defined as regulation whose costs are higher than the benefits obtained. Do our regulations cost us more than they benefit us? Certainly there are more minorities and women in business today than there were before the law required such hiring. The air is cleaner, the water is less contaminated. This has been accomplished at a tremendous cost, in some cases at the cost of a total business. In many cases the cost has been too high because the standards have been too high. I do not believe our drugs are appreciably safer now nor do I believe more effective drugs have been produced as a result of the 1962 drug law revision.

I believe there is a practical limit beyond which we need not go to effect the protection of lives, the preservation of the physical environment, the continued maintenance of our economic health. That limit is not the elimination of all traces of foreign substances from the air or the water, or the protection of every living individual plant or animal. "Pure" air and "clean" water and "safe" chemicals are defined by whatever way we want to define them. Pure and clean and safe mean to me that we can live in an environment that does not harm us. It does not mean the environment from which we have removed everything that is thought *could* harm us.

No organization can be creative if it is not free, free to think and free to do. The more an organization is regulated, the more its activities must become restricted to satisfy these regulations, and the more its financial resources must be diverted from risk-taking activities. The path of creativity is littered with regulatory obstacles. It is no wonder, then, that business administrators find it difficult to commit themselves to flying when they know their flights will be limited by the number of regulatory millstones hanging from their necks.

4 . . . wherein an attempt is made to remove all risk from our lives, and there is no difference between human beings and other animals, and doubt is raised about the advisability of eliminating risk

Risk is the chance for harm or loss. Risk is bad. Everyone should avoid risk. If people do not avoid risk on their own initiatives then someone should force them to be cautious. Since individuals seem not to appreciate the badness of risk and, in some cases seem even to like it, the government, as the great protector, must see to it that all risks are eliminated. The government takes this responsibility so seriously that it even protects against risks that do not exist. It has had so much experience with risk that it sees risks where no one else sees them. The idea of protection is so attractive that when all else fails, agencies manufacture risks to protect the public against.

A benefit is anything that is for the good of an individual or society. Benefit is good. Everyone should strive for benefit. If people do not strive for benefit someone else will decide what a benefit is and force it upon them. The government has done its part by preventing risk. The elimination of risk is the benefit.

James Russell Lowell, in *A Fable for Critics* said:

> For there's nothing we read of
> in torture's inventions
> Like a well-meaning dunce,
> with the best of intentions.[1]

Far be it from me to suggest that those responsible for some of the regulations designed to protect us are dunces. However, be they dunce or genius the results of their actions are the same.

The desire to have an obvious danger removed is part of the instinct for self-preservation. Consequently when someone in a position of authority or someone believed rightly or wrongly to be an expert warns of danger, the almost universal reaction by the public is one of acceptance of the restriction that keeps us safe. One of the factors contributing to the overly fearful reaction to potential hazards is that we are living in an age of instant communication. In the past, postulations could be presented, analyzed, and accepted or rejected before decisions were made. Now, the postulations become public and they are widely broadcast immediately. "What if" questions are now converted in the minds of the public into "It will" answers. An indirect result of this is that with general public knowledge more pressure is brought to bear on those responsible for making decisions. Now the mere fact that someone has raised a question results in forced, often premature, and often wrong decisions.

Examples are legion. In 1973, based on a report from a single physician, the United States Consumer Products Safety Division banned spray adhesives because they caused birth defects. Six months later the ban was lifted after further studies showed that the spray was safe. However, the initial report had received instant publicity. In February of 1976 an article appeared in *Science* magazine reporting that at least eight women who had been exposed to the spray adhesive elected to have abortions because of the fear of having a deformed child. Other women asked for chromosomal screening as a result of this fear.

The accident at the Three Mile Island nuclear plant was one of the most thoroughly reported incidents of all time. Every utterance of any individual, official or non-official, connected with the plant or living in the vicinity was conscientiously passed along. Unfortunately, there was no discrimination used in interpreting or analyzing these opinions. One of the most tense of the tense situations reported was the possibility of a hydrogen explosion that would blow up the plant. For at least three days the progress of the "hydrogen bubble" was reported from hour to hour. Suggestions were made that the entire area be evacuated because of the possibility that the plant might blow up. In early May the director of the Nuclear Regulatory Commission reported that there never was any danger of an explosion. It took thirty-six hours to figure this out, and by that time the entire world had been assuming there was a definite possibility of tragedy. The director told a subcommittee of the Advisory Committee on Reactor Safeguards: "The amount of concern was entirely undeserved. There never was any danger of a hydrogen explosion in that bubble. It was a regrettable error."

In late October of 1979 nuclear regulatory officials summoned nuclear industry representatives on forty-eight hours notice to a meeting in Washington to discuss a problem that could require power reductions at every nuclear plant in the country. A two-year agency research study had concluded that there was the danger of cooling tubes being blocked and bursting from the increased pressure. The very day after the meeting, the NRC commissioners were hastily briefed by the NRC staff, who had concluded that the suspected problem occurred only in the laboratory tests and did not occur under operating conditions. However, the initial results had been publicized and there were widespread demands for the shutdown of all nuclear plants. Lamented the Chairman of the NRC: "Today's exercise is one of the penalties one pays for the benefits, presumed, of promptly putting new information into the public record." The final report of the House Science and Technology Committee published a year after the accident concluded that there was no great threat involved. The chairman of the committee said: "Three Mile Island was a serious accident but not a serious threat to human health. The greatest harm from a Three Mile Island accident was its severe emotional impact on an ill-informed and easily frightened public, especially near the plant."

In October of 1979 a report prepared by a presidential commission investigating the accident was released. Some of the conclusions reached by the committee after more than a year of retrospective examination of the facts are interesting in view of the hysteria that prevailed during the day-to-day reporting at the time of the accident. The committee said:

> Fortunately, in this case, the radiation doses were so low that we conclude that the overall health effect will be minimal. There will either be no case of cancer or the number of cases will be so small that it will never be possible to detect them. The same conclusion applies to the other possible health effects.
>
> We conclude that the most serious health effect of the accident was severe mental stress, which was short-lived.
>
> Issues that had to be examined were whether a chemical (hydrogen) or steam explosion could have ruptured the reaction vessel and containment building, and whether extremely hot molten fuel could have caused damage to the containment. The danger was never—and could *not* have been—that of a *nuclear* explosion (bomb). [Emphasis in original report]
>
> On the basis of present scientific knowledge, the radiation doses received by the general population as a result of exposure to radio-

activity released during the accident were so small that there will be no detectable additional cases of cancer, developmental abnormalities, or genetic ill health as a consequence of the accident at TMI.

The great concern about a potential hydrogen explosion inside the TMI-2 reactor came over the weekend. That it was a groundless fear, an unfortunate error, never penetrated the public consciousness afterward, partly because the NRC made no effort to inform the public that it had erred.[2]

In another example, flammable children's wear was prohibited. A chemical, Tris, was discovered to be a means of fireproofing fabric. It was accepted as the universal flameproofing material. It was then found that Tris caused a form of cancer when fed to rats in large doses. It was immediately banned as a flame retardent. A substitute was discovered, put into general use, and, within a year, was also banned for the same reason.

In the drug field many examples could be given where great benefits outweigh obvious risks. An antibiotic, a specific cure for Rocky Mountain Spotted Fever, causes fatal anemia in one of 10,000 persons who take it. The fever kills 8 out of 10 people who have the disease. Obviously the antibiotic should be used. Yet, in this atmosphere, it would be difficult today to get such a drug into clinical trial.

There is something to be said for a system that keeps dangerous drugs off the market. But the same system, if not used properly, also makes highly effective drugs unavailable. If aspirin and penicillin were discovered today they probably would never be accepted for clinical trial, either because there is no evidence in animals that the drug is effective, as is the case of aspirin for headaches, or because the drug is highly toxic in some species of animals, as is the case with penicillin.

As opposed to protection by preventing action from taking place is the equally extravagant forcing of protection on consumers. Despite controversial studies as to its effectiveness, owners of automobiles might have been required to pay for airbags by 1983 if a proposed law had been passed by Congress. The Ford Motor Company has estimated that the airbag would cost buyers $825. However, if the bag is once released the General Accounting Office says that the replacement cost is $2,100. Critics say it is worthless and even dangerous in other than frontal crashes. The National Highway Traffic Safety Administration says it will prevent about 9,000 highway deaths a year. Again, the concept seems admirable—to protect the public from themselves, even when the public thinks the protection is unnecessary and the method useless. However, in the closing days of the lameduck 96th Congress an attempt to mandate the use of the bag was defeated by only three votes, 189 to 186.

At one point it was illegal not to wear seat belts. To assure that all drivers wore seat belts, it was made a legal requirement that a device be built into all automobiles that would prevent the car from starting if the belts were not fastened. The public outcry was so great that the requirement was eliminated. Like prohibition, it was a noble experiment, and, like prohibition, it was a law that could not be enforced.

The National Traffic and Motor Vehicle Safety Act required the registration with the Federal government of all retreaded tires "to enable the government to move quickly in the event a recall is necessary because of a product defect." The Chairman of the Senate Consumer Subcommittee of the Committee on Commerce, Science, and Transportation reported that 63 million tires had been registered since 1971 at a cost of $3 million per year, a total of $24 million. During that entire period only eight tires were recalled. Three million dollars per tire seems like an unreasonable price to pay for control. The government, obviously, did not pay for the control. It was paid for by the general public, in increased cost of tires.

The automobile industry was given a timetable to reduce exhaust emissions. The catalytic converter was the result. It did reduce some emissions, and, in spite of its expense, was thought to be a success. Now we discover that excessive amounts of sulphur dioxide are emitted that might be more harmful than the emission the converter was developed to eliminate. The converter requires a significant percentage of both the platinum and the stainless steel available in this country. With a more realistic timetable after the problem was upon us, or with better planning before the problem became a reality, a better system could have been developed.

Another example of an ill-conceived approach to safety is that of OSHA's attempt to eliminate exposure to carcinogens, certainly a desirable objective. One of the first actions taken by OSHA was to employ a consulting company to survey the scientific literature. If they found two references indicating positive carcinogenic results from a product the product was listed as a carcinogen. If negative results were found for the same products they were discounted. Back in 1960 I made my own somewhat superficial literature survey as part of a presentation to a Congressional committee. A partial list of products reported as carcinogens in reputable scientific journals includes coffee, tea, milk, cream, cocoa, claret wine, caffeine, sulfonamides, fat, cholesterol, vitamins, extracts of human tissues, eggs, sugar, various plastics and synthetic fibers, detergents, common solvents, and metals. Estrogens have been shown to cause cancer in animals, and the following foods contain estrogens: normal beef, milk, soybeans, corn, lettuce, alfalfa, clover, wheat, oats, lobster, beets, potatoes, and honey. Sunlight is a well known cancer producer. At least thirteen trace elements found in animal bodies are carcinogenic in animals. But some of these same elements are essential to life.

Even Scotch whiskey and beer have become suspect. In the fall of 1979, mainly as a result of an "investigative report" by a television station, national publicity was given to the fact that both Scotch whiskey and beer contain a more potent nitrosamine than that found in cured bacon. It did not make too much difference to the publicizers that the original work on beer, done in Germany, had been published over a year earlier, and that a scientific committee to study the problem had been operating in the United States since that time. It was believed that the nitrosamine is introduced into the product during the processing of the barley used in the fermentation process. By the time the publicized "danger" had been spread across the country, the problem, such as it was, had been solved, nitrosamine levels were well within safe limits, and the beer and Scotch drinkers who seemed to be singularly undisturbed by the whole affair continued drinking apace. The protectors of the public could take satisfaction in the belief that they had eliminated the threat of cancer, leaving only the natural threats of alcoholism, cirrhosis of the liver, accidents due to inebriation, and deaths due to the combination of alcohol with other drugs, just as God had intended.

If we attempt to remove anything that has been shown to cause cancer under any conditions in animals we would be forced to spend our lives naked and hungry in some dark cave. And even then we would not escape background radiation.

Anything that becomes a part of a food is classed as a food additive. Therefore, when Monsanto discovered that acrylonitrile, a component of its plastic bottle, migrated into the beverages in the bottle at an average of ten parts per billion after six months at room temperature, the Delaney amendment required FDA to ban the bottle, since acrylonitrile had been classed as a carcinogen in animals. Tests run by Monsanto showed that at the lowest level of feeding to animals that causes harm, it would require that a person drink 3,000 bottles of soft drinks every day for life to get the same results. This decision was appealed and the Appeals Court ruled that FDA could not classify a substance as a food additive merely because it came in contact with food. The Court ruled that, to be classified as a food additive, the substance must be shown to be present in food in more than insignificant amounts.

Of course all the poisons that occur in food are not additives. Some, in fact most, occur naturally. For example there are at least 150 separate chemicals in a potato, among them being arsenic and a toxic alkaloid. Corn and peanuts contain aflotoxins, extremely toxic materials that also cause cancer in animals. The aroma in coffee is caused by 42 chemicals. Cheese contains amines that can increase blood pressure. Shrimp has a high arsenic content. Because the chemicals are "natural" they are accepted as safe. However, they are no less toxic than additives. They are safe because they are consumed at a level below that at which side effects occur. It seems illogical to accept a chemical because it is natural and oppose another chemical, possibly less toxic, because it is added to a food to perform some useful function.

Over the years the concept of zero tolerance or zero level has developed. If a dangerous chemical was used as an insecticide or as a growth promotant for animals, the food that was ultimately consumed had to be free of the chemical, that is, have a zero residue. Today the same concept is being applied to all kinds of ecological problems.

As a practical matter, "zero" has to be defined. In the past, as now, "zero" must be limited to the sensitivity of the instruments measuring contamination. For example, if an analytical method could detect one part per million of a substance but was not sensitive enough to detect 0.75 parts per million, then anything containing 0.75 parts would be reported as zero since the test results would show no contaminant present. More and more sensitive instruments have been developed for detecting products in the human body, in food, in water, in the air. It was not too many years ago that a concentration of anything below one part per million was considered "zero," and instruments that measured to that accuracy were considered fantastically sensitive. Now, it is routine to measure parts per billion or parts per trillion. It is ironic that advances in the technology of measurement have created a major problem for other areas of technology. What had in the past been considered zero is now no longer zero. It is two parts per billion or five parts per trillion or some equally small number. If, then, the thinking prevails that says that anything that is measurable is dangerous, what was safe ten years ago is now forbidden by law, even though nothing but the ability to measure has changed.

How small is one part per trillion?

> One part per trillion is one inch compared to the distance of 32 trips to the moon.
>
> One minute in the two million years that humans have been present on earth.
>
> One penny in ten billion dollars.
>
> One large mouthful of food when compared with the food one million people eat in their entire lives.
>
> One drop of vermouth in 80 million fifths of gin, a very dry martini indeed.

I think it is obvious that, as the ability to make more and more sensitive measurements increases, the thinking about the significance of the measurements must also change. When measurements were relatively crude compared to what is available now anything that was measured was obviously a significant amount. Now, with the exception of some very rare instances such as the presence of toxins, I think it is ridiculous to think of exposures, usually very short term and intermittent exposures, to parts per trillion of anything as being dangerous.

It is hard to argue against the logic that says that if the concentration of an undesirable material can be reduced from ten parts per million to one part per million or ten parts per billion, then it should also be reduced to zero. Possibly it can be totally eliminated, but at what cost? It is an almost inviolable rule that the cost of reducing the concentration of chemicals in the atmosphere increases geometrically as the concentration approaches zero. The technical difficulties are not appreciated. Thus, the public belief that, if a pollutant can be reduced to half its original concentration it is only the stubbornness of the organization causing the pollution that prevents its being cut in half again and again until practical zero is reached. It is also difficult to argue against the logic of reaching zero if some definite benefit is shown when, for example, a contaminant is reduced from 100 to 50.

Society spends $700 million a year to reduce carbon monoxide auto emission to 15 grams per mile. This saves thirty thousand person years, or prolongs the life of thirty thousand people an average of one year at a cost of $23,000 per person year. The 1981 standards have been lowered to 3.4 grams per mile. General Motors estimates it will cost $100 million more to meet this standard, while producing only twenty additional person years at a cost of $5 million per person year.

Some sense of values must be exercised. When the U.S. highway speed limit for automobiles was reduced to 55 miles per hour, the Transportation Department reported that highway accident deaths declined from 54,052 in 1973 before the reduction to 45,196 the year after the reduction. I can tell you how to save an additional 50,145 lives, the number of motorists killed in 1978—simply reduce the speed limit to zero. That proposal makes as much sense as some of the suggestions made to reduce chemical exposure to zero. In fact it makes more sense, since it is known for sure that 50,000 lives are lost as a result of automobile accidents, and it is only postulated that lives will be lost due to chemicals.

The fallacy in this thinking is obvious. The risk has been removed, but so has the benefit.

Again, we come back to the concept of being risk-free. "If we err, it is better to err on the side of safety." For some, then, safety dictates eliminating everything that is not considered "natural." If it can be measured it is something tangible and should be eliminated. Unfortunately, such is the attitude of both public and administrative bodies and regulatory agencies. The obstacles that the restrictions resulting from such thinking present to the technological industry are obvious and formidable. One of the hazards of over-emphasizing a potential danger is the possibility that warnings will become so commonplace that they will not be taken seriously. The next step is a backlash against them, and the ultimate in rejection is ridicule. The "carcinogen of the week" is already a common expression.

There was formed at St. Johns, Newfoundland a new protest group called Cod Peace whose objective it is to protect the defenseless codfish from the savage harp seal. The seal eats about ten pounds of cod per day. It is not known if the seal gets its ten pounds from two five-pound cod or takes bites from a number of smaller fish. "Imagine the environmental damage," solemnly proclaimed the group's founder, "that would result if a herd of harps decimated a school of cod, leaving a sticky oil to be washed onto the beaches."

A newspaper column reported on the reaction of caribou to the Alaskan pipeline. During the construction of the line special gates were built at the migrating points of caribou. This necessitated putting the pipe underground at considerable added expense. Special precautions were taken to prevent its heat from melting the snow and the ground above the pipe. The observer of the caribou reports that the caribou actually seem to like the pipe above the ground. The column concludes: "They sleep under it, play under it, leap over it, enjoy its friendly warmth. What can you do with animals like that?"[3]

Satirical suggestions for the regulation of nature are being made. A report from the San Francisco Bay area indicates that trees may cause more air pollution than automobiles. The Appalachian forest caused an ozone alert in Washington, D.C. Radiation from ores in the earth and radiation from the sun are more widespread than anything resulting from nuclear plants.

Human-caused catastrophes pale into insignificance when compared with the devastation caused in the past and still being caused by nature. Bubonic plague wiped out one-fourth of the population of Europe. In 1918-19 twenty million deaths resulted from influenza. The pollution caused by volcanoes overshadows all the human-produced pollution combined. The recent eruption of Mt. St. Helens is dramatic immediate evidence of that. The Endangered Species Act is designed to protect one or two hundred species from extinction. It is estimated that about 500 million animal species came into being through the ages but became extinct with no stimulus from people.

The agencies do nothing to emphasize the importance of the situation when they publish information that seems to trivialize important warnings. OSHA has published a list of "hazards of arts and crafts." They include ceramics, dyeing, fiber arts, jewelry, metal sculpture, plastic sculpture, stained glass, and woodworking. For example, under ceramics, they list as dangerous asbestos, silica, glaze components (lead, barium, lithium), colorants (copper, manganese, nickel, chromates, cadmium, antimony, uranium), kiln gases (sulphur dioxide, carbon monoxide, fluorine, chlorine, nitrogen oxides). Under plastic sculpture we are warned against styrene, methyl-methacrylate, diisocyanates, formaldehyde, organic peroxides, aliphatic amines, solvents (acetone, methylene chloride, ethylene dichloride, lacquer thinners, etc.), carbon monoxide, hydrogen chloride, hydrogen cyanide, asbestos, silica, and fiber glass.

It somehow boggles the mind to think of artists' studios sometime coming under OSHA restrictions. It is even more mind boggling to think of how the artist would react.

In a single news release the National Cancer Institute announced that four widely-used drugs produced cancer in animals. These are Methapyriline, a drug used in sleep aids, Reserpine, a product widely used in hypertension, Seleniumsulfide, used in anti-dandruff shampoos, and Disulfiram, a fungicide and anti-alcohol drug.

In June of 1979 the FDA requested Upjohn to revise the warning label for its drug, Depo-Provera. In a ten-year monkey study two of twelve monkeys had developed uterine cancer after receiving large doses of the drug over long periods of time. Depo-Provera is already on the market approved for two uses by the FDA. What are the uses? It is used as a treatment for endometrial cancer, the same type of cancer appearing in the monkeys, and for the treatment of kidney cancer.

Not only is the idea of a risk-free existence being proposed as a possibility for human beings but it seems now that the same concept is being applied to plants and animals.

In 1973 Congress passed the Endangered Species Act that says that nothing can be done that would jeopardize the existence of endangered or threatened species, or destroy habitat deemed critical to their survival. In 1975 a small fish, the snail darter, was placed on the endangered species list, and the Little Tennessee river, the only known habitat of the darter, was declared critical. Sometime before the listing, construction of the Tellico Dam, a $120 million project of the Tennessee Valley Authority, had been started. Completion of the dam would have eliminated the habitat of the darter and thus destroyed the last surviving members of the species. With about $103 million of the $120 million already spent on construction, a lawsuit based on the Endangered Species Act halted further building. For three years the uncompleted dam was the center of controversy in the courts. Whether or not to assure the continued existence of a small fish was placed in the hands of the Supreme Court. And the Supreme Court did decide. It decided that, under the law, the snail darter had to be protected "at whatever the cost," in the words of Chief Justice Burger. He further stated that, since Congress had been so specific in its wording and had left no room for qualification, the court had no alternative. The decision could be interpreted as saying that, if Congress passed a foolish law, the Supreme Court in this case could not rescue them. The foolishness was compounded by the fact that Congress continued to appropriate money for the Tellico Dam, even after the presence of the snail darter was known to them, even after the probable objections to its completion were evident, and even after the Supreme Court decision had been announced.

But the circus was not yet over. Senators whose constituents would be affected if the Tellico Dam was not completed were successful in securing an amendment to the Endangered Species Act of 1973 creating a review committee with the power to grant exemptions from the act under certain circumstances. In January of 1979 a request was made to the committee to allow the Tellico Dam project to continue under an exemption. The committee established as a means to allow the dam project to continue voted unanimously not to grant the exemption.

The fury of a woman scorned is as nothing compared to the fury of politicians seeing available funds not being appropriated to their states. With typical, if not admirable persistence, the politicians continued their battle for the dam. So it was that, in late 1979, in spite of the fact that it overrides the Endangered Species Act, in spite of the unanimously adverse judgment of a special review committee, in spite of the fact that both the Secretary of the Interior and the Chairman of the Council of Economic Advisors had stated that, even if completed at an additional cost of $19 million, it would not be cost effective, and in view of past Senate negative votes, both the House and the Senate voted an exemption to allow the dam to be completed.

Two examples of conflicting national objectives provide an indication of our national priorities. In the late 1970s the solving of our energy crisis was certainly one of the critical problems facing the nation. But even such a national crisis had to take second place to a real or imagined insult to the environment. In Redwash, Utah, Chevron Oil wanted to start an oil drilling program after the land had been covered with snow. The Bureau of Land Management required an archaeological check for Indian artifacts before drilling began. Attempts were made to remove the snow with hot water. The Bureau of Land Management prevented this because of the danger of erosion. When flame throwers were proposed they were eliminated because they might destroy loco weed that might be an endangered species. The only solution was a delay of months until the snow melted, with a consequent delay, also, in oil production. Neither arrowheads nor endangered loco weed was found.

The other example was in the same category. A wildcatter in Wyoming had to delay oil exploration for three months because "noise from around-the-clock drilling operations tend to suppress sage grouse strutting activities." Strutting is part of the mating ceremony of the grouse so, when drilling overlaps the mating season, drilling in the area must cease.

Further examples that, on the surface, appear ridiculous but that are required by a literal interpretation of the law are numerous. In Ohio, a bill came before the State Legislature that would allow Indians of that state to pluck the feathers of non-game birds for their headdresses. However, before the plucking took place the Indian would have to prove he or she was a real Indian, and a feather plucking permit from the Division of Wild Life would be required. A permit would not allow birds to be killed, but would give the Indian only the right to collect feathers from birds found dead. A final stipulation was that the feathers could not be given to another Indian unless the original feather plucker died and bequeathed them to a new owner.

Even the Supreme Court of the United States believed the collection of feathers to be important enough for them to decide a case at the highest level of law. In Colorado, several dealers who sell and appraise American Indian artifacts questioned the fact that they were not allowed to deal in eagle feathers. The Supreme Court in an unanimous decision ruled that eagle feathers, even though collected years ago, could not be sold. The Court said: "There is no sure means by which to determine the age of bird feathers; feathers recently taken can easily be passed off as having been obtained years ago."

The Endangered Species Act can be used as a two-edged sword, with environmentalists themselves feeling the frustrations when the law is used against them. In an effort to reduce the municipal sewage discharge into the ocean from the city of Los Angeles the EPA mandated a project that would have required the discharge from the secondary treatment to have been carried five miles out into the Pacific Ocean by means of an underwater pipe. The city stated that the new plant would cost about $250 million to build and another $40 million to operate, without improving the environment. A public interest law firm promptly filed suit, charging that removing the sewage would harm three endangered species—the gray whale, the brown pelican, and the El Segundo blue butterfly. The reasoning was that eliminating the sewage would remove nutrients necessary to support the large population of fish around the sewage outlets. The fish, in turn, were a source of food for the gray whale and the brown pelican, and, therefore, anything that interfered with the fish also interfered with endangered species. The butterfly lives in a field near the plant site. Its habitat would have been destroyed if the new system were put into effect. A Federal judge in Los Angeles agreed with the public interest group, and barred the Environmental Protection Agency from proceeding with the project until EPA had complied with the procedures of the Endangered Species Act.

In May of 1977 two students removed two dolphins being held for experimental purposes at the University of Hawaii Institute of Marine Biology, and released them into the ocean. When the students were brought to trial the defense was based on the fact that dolphins were intelligent beings, and, as part of a "declaration of dolphin rights," said: "In the spirit that moved lawmakers to enfranchise first men with property, then men free and white, and finally women, we plead with today's lawmakers to treat generously that intelligence of the sea . . . ." The students were found guilty of grand theft.

A similar situation arose in California when a female gorilla named Koko was loaned by the San Francisco Zoo to a Stanford University graduate student who taught the gorilla to use sign language. When the zoo demanded the gorilla back, several lawyers objected, stating that Koko now had human abilities, and should have the same constitutional rights as a human being. Fortunately, the case never came to court. The University solved the problem by setting up a foundation to purchase the animal from the zoo.

The 1973 Endangered Species Act was passed with the best of intentions—to prevent the elimination of any species from this earth. What is so wrong with the elimination of a species? After all we have about a million animal species and a quarter of a million plant species as fellow travelers on our spaceship earth. It has been estimated that an additional 500 million animal species have come into being and disappeared since the earth began. So why try to preserve a few species that only a few hundred people have ever seen, and not many more than that have ever heard about?

Different people would give different answers. The scientist would say that it is important to preserve every possible representative of the gene pool. The ecologist would say that every species, regardless of type and number is part of the chain of life, and unknown and unfortunate consequences might result from the interruption of a seemingly inconsequential part of that chain. The ethical and moral values of some people would lead them to say that every living thing, plant or animal, has the same rights as human beings. Whatever the reason for protecting plants and animals the result has been another restriction for some of the activities of business.

The original Endangered Species Act listed 109 species that required protection. When most people think of endangered species the bald eagle, the whooping crane, or the grizzly bear might come to mind. Certainly no one could object to such protection. But, since 1973, 24,000 plant and animal species have been suggested for the endangered list.

The Tellico Dam is not the only project being held up because of action taken as a result of the Endangered Species Act. On the banks of the St. John River in Maine there grows a plant called the furbish lousewort. This is only one of a family of louseworts, so called because farmers once believed that sheep and cattle grazing on these plants became infected with lice. The banks of the St. John is, as far as anyone knows, the only place where the furbish lousewort grows. Therefore, if a proposed dam, the Dickey-Lincoln Dam, is built, it will destroy the habitat of the lousewort causing it to disappear. In view of the straightforward and uncompromising wording of the law it is likely that the lousewort will be the victor in the coming court trial over the construction of the dam.

The Duck River Dam in Tennessee is in trouble because, in the river proposed for damming, live the tan riffle shell, the orange-footed pimpleback, and the Cumberland monkeyface. I am not making this up. They are all mussels.

In a ninety-foot length of drain pipe leading from an abandoned bathhouse three miles west of Socorro, New Mexico, live the last known 2,500 Socorro Isopods. The Isopod, a crustacean relative of the common cowbug, is in danger of extinction because the pipe is occasionally flushed and, in addition, the organisms are cannibalistic. If this crustacean is declared an endangered species the government would take custody of the "critical habitat," the ninety feet of rusty pipe. How the problem of cannibalizing will be taken care of I do not know.

Work on the Flaming Gorge Dam on the Colorado River was delayed because of a threat to the Colorado squawfish. Only months before there had been a government campaign to eliminate the squawfish because it was destroying trout and other food fish in the streams.

An example of the necessity to make a decision between the life of an animal and the life of a human being is already here. A virus-caused disease, hepatitis B, is widespread around the world. It is estimated that 150 million people are carriers of the virus. In the United States alone 15,000 cases were reported in 1976. The Center for Disease Control estimates the real incidence here is about 150,000, with 1,500 deaths resulting. A vaccine, presently under development, would prevent the disease. However, aside from humans, chimpanzees are the only species suitable for testing the safety of the product. Each chimp can be used only once. Consequently, if the vaccine becomes a production item a large number of animals would be required. The pharmaceutical house developing the product applied for a permit to import 125 chimpanzees from Sierra Leone. The plan is to use the animals once for testing and then transfer them to a breeding colony.

The application was vigorously opposed by conservationists, and the U.S. Fish and Wildlife Service has asked for more information from the manufacturer. The manufacturer contends that, not only would the availability of the chimps make the vaccine possible, but the breeding colonies proposed might be the one way to preserve this endangered species. The environmentalists have objected to the method by which the chimps are captured, and point out that chimps hold a special place in the animal kingdom, since they not only have an advanced social structure but now can be taught to communicate with humans by means of sign language.

It is a dramatic choice—a choice between humans and their closest related species. If chimps are not used there will be no vaccine, and the deaths of human beings around the world will not be prevented. If chimps are used some chimps will probably die as the result of their capture. But these are the choices that are being forced upon us now.

Now attempts are being made to export U.S. values, to use our economic strength to force foreign countries to conform to our ideas of environmental control. In 1977 the National Resource Defense Council filed a suit against the Eximbank to force it to provide protection of foreign environments that might be affected by projects supported by the bank's loans. The Eximbank is the export-import bank of the United States. In 1976 the Eximbank credits and guarantees supported nearly $12 billion in United States sales to 157 markets. The bank estimated that these sales were related to 500,000 American jobs. One of the projects affected by this action was in the Republic of Gabon, where the government had been attempting to build a 440-mile railroad across the country, deemed essential for the country's development. The NRDC suit contended that the railroad might endanger the habitats of gorillas, crocodiles, buffalo, and elephants.

I suppose the ultimate in recognizing the right to be risk-free was reached when, in March of 1978, a Los Alamos scientific laboratory worker was awarded $75,000 as the result of a claim of disability from neurosis. The award was based solely on the fact that the man's occupational environment made him nervous. No showing of any physical harm was made or claimed, nor was there any claim that working conditions were unsafe. He simply worried that he was going to die of cancer.

All this is done in the name of protecting human beings, animals, and plants from risk. But there can never be a risk-free environment. I think, rather than trying to eliminate risk from our lives, we should, instead, concentrate on learning what risks are challenging us, and what the consequences of accepting or rejecting these risks are. A risk, by definition, is not an inevitably bad occurrence. A risk implies that there is a chance that a bad effect may or may not take place. In the present atmosphere, risk has been interpreted as meaning that the result of the action will always be undesirable. What we should be asking is what benefit might result from the risk we are taking, what is the chance that the bad effect will occur, and what is the cost of the effect. In the oft-repeated but seldom heeded advice, "weigh the benefits against the costs."

Cost-benefit analysis it is true, is a difficult exercise and a limited tool at the present time. The cost in dollars to meet regulatory standards can be calculated with some accuracy. The benefits are not so easy to evaluate. As a result, there has been an attack even on the concept of having to justify with numbers any regulation whose purpose is ostensibly to protect the public or the environment. Proponents of more and stricter regulations argue that it is not just difficult but that it is inappropriate as well.

It is, I suppose, a natural position to take when you are totally convinced you are right, and that what you are doing is for the public good. Why spend all the time and money to simply prove what you know is right already? Let industry prove its position. After all, they are the ones who have to pay the bill anyway. But industry does not pay the bill. Industry's customers eventually pay the bill. And the bill is not just the dollars spent in meeting regulatory requirements. That is only part of it. That is the part that is seen in higher prices and in inflation. The unseen part is the cost of not doing creative things that could be done. The unseen part is the lack of availability of drugs that could be made available, or the new processes that would result in decrease of consumer prices. The unseen part is the high cost of manufacture and low productivity that make it impossible for United States industry to compete with foreign products, that result in our negative balance of trade, that result in the devaluation of the dollar abroad, and that, in general, contribute to our whole unstable economic situation.

Why shouldn't justification of regulations be required, particularly if they are to cost the American public, not American industry, billions of dollars? It is granted that estimating the costs, benefits, and risks is difficult. But some start is being made, and the mere existence of situations that cause bureaucracies problems should not be the excuse for allowing regulatory agencies to operate in comfort at the expense of the public.

One effort to bring some logic into the determination of what is an acceptable risk has resulted in "risk accounting." Risk accounting, as the term implies, means taking into account the risk involved in every operation that is necessary to achieve an objective. For example, in obtaining power from solar energy more is involved than just allowing the sun to strike an absorbing surface. Machinery and equipment must be built. Materials must be mined or manufactured from which the equipment is to be made. To each operation, then, a risk is attached, and the total risk of obtaining solar energy is accounted for.

Dr. Herbert Inhaber, of the Canadian Atomic Energy Control Board, was one of the first to attempt this assessment.[4] He applied it to an analysis of the risks involved in obtaining a specified amount of energy from different energy sources. Using the following sources he had estimated the range of deaths caused in producing an energy output of 10 gigawatts per year.

Coal 50-1600
Oil 20-1400
Wind 120-230
Solar 80-90
Uranium 2½-15
Natural Gas 1-4

What is significant is not the wide variation in the individual categories. It is the approach that has been taken, a true evaluation of not just a risk of operating a certain procedure but of building it and making it possible to operate. When we evaluate the risk of energy from coal we must start with the number of miners killed to produce the coal that allows the power plants to operate. The same reasoning will be applied to the refining of uranium. However, since only a very small fraction of uranium is needed compared with the amount of coal to produce an equal amount of energy the risk of accident is greater for the coal miner than for the uranium producer. In 1978 there were 100 coal miners killed in the United States, down substantially from the 300 killed in 1969. If we take this approach a whole new basis for evaluation becomes available. Even the peaceful-looking windmill becomes dangerous, since it produces a comparatively small amount of energy and therefore requires considerably more work and therefore more risk to make it equivalent to a ton of coal.

Would you believe that there is more danger from the radioactivity from a coal-fired power generating plant than from a nuclear power generating plant? Preliminary data were published by EPA in their report, *Radiological Impact Caused by Emissions of Radionuclides into Air in the United States.* According to the report, radionuclides (radioactive atoms) are disseminated into the atmosphere where they are breathed or swallowed. Coal contains radionuclides of radon, uranium, actinium, and thorium, as well as potassium-40. When coal is burned most of the radioactivity remains in the residue as slag or ash. However, some is vented into the atmosphere as fine ash. According to the EPA, from 0.0004 to 1.5 fatal cancers per year can be expected from each of the 250 existing coal-fired power stations in the United States. By contrast, each of the existing 69 nuclear power generating plants is estimated to cause

0.001 fatal cancers per year.

New technology, as well as removal of the plants from urban areas, change these figures. On a direct comparison at suburban sites between coal and nuclear plants, a boiling water reactor nuclear plant produces 0.0013 fatal cancers per year, and a pressurized water reactor is responsible for 0.0009 cancers per year. However, existing coal-fired plants produce 0.10 fatalities per year, while coal-fired plants using new technology produce 0.017 deaths.

Lord Rothschild, former director of the British government's Central Advisory Council for Science and Technology, in an address on BBC television in November, 1978, had some perceptive things to say about risk. He says: "There is no point in getting into a panic about the risks of life until you have *compared* the risks which worry you and those that don't—but perhaps should. Comparisons, far from being odious, are the best antidote to panic. What we need, therefore, is a list of index of risks and some guidance as to when to flap and when not." He says that "Even a *virtuous* life has risks" and he quotes what he alleges is a Chinese proverb: "The couple who goes to bed early to save candles ends up with twins."

Professor Bernard L. Cohen of the University of Pittsburgh has taken the interesting approach of estimating what influence the exposure to various risks, including socioeconomic situations, will have on individual life expectancies. Writing in the June, 1979, issue of *Health Physics* he examined over fifty different life situations. Using the available information such as mortality rates and exposure to external factors, the loss of life expectancy in days was calculated, with some surprising results. For example, some of his conclusions are:

Cause	**Days of Life Expectancy Lost**
Being unmarried—male	3,500
Cigarette smoking—male	2,250
Heart Disease	2,100
Being unmarried—female	1,600
Being 30% overweight	1,300
Being a coal miner	1,100
Cancer	980
Alcohol (U.S. average)	130
Job with radiation exposure	40
Illicit drugs (U.S. average)	18

Cause	Days of Life Expectancy Lost
Natural radiation	8
Medical X-rays	6
Coffee	6
Oral contraceptives	5
Diet drinks	2
Radiation accidents (U.C.S.)	2
Radiation accidents (Rasmussen)	0.02
Radiation from nuclear industry	0.02

The radiation accident figures are calculated using two different bases. The higher risk is from data of the Union of Concerned Scientists, the lower is from a report written for the Nuclear Regulatory Commission by Norman Rasmussen of MIT. All three radiation calculations assume that all United States power comes from nuclear sources. In addition, Professor Cohen's data concludes that the Pap test increases life expectancy by four days, smoke alarms by ten days, airbags in cars by fifty days, and mobile coronary care units by 125 days.

The loss of life expectancy data bring forth some interesting exercises in priority setting. If the primary objective is to increase the life span of our citizens and eliminate those things that might shorten a life, then much more attention should be given to marrying off all the unmarried males and females then is given to eliminating nuclear power. This conjures up some frightening possibilities for bureaucratic action.

In a similar vein Dr. Richard Wilson, Professor of Physics at Harvard Universtiy, has calculated what degrees of various activities are necessary to produce the same risks. Writing in the February 1979 *Technology Review,* Professor Wilson estimates that each of the following actions will increase the chance of death in any one year by one in one million:

Smoking 1.4 cigarettes
Drinking one-half liter of wine
Living two days in New York or Boston
Living two months in Denver on vacation from New York
Living two months in an average stone or brick building
One chest X-ray taken in a good hospital
Eating forty tablespoons of peanut butter
Drinking Miami drinking water for one year
Drinking thirty 12 oz. cans of diet cola

Living five years at the site boundary of a typical nuclear power plant in the open
Drinking one thousand 24 oz. soft drinks from a banned plastic bottle
Living twenty years near a PVC plant
Living one hundred and fifty years within twenty miles of a nuclear power plant
Eating one hundred charcoal-broiled steaks
Risk of accident by living within five miles of a nuclear reactor for fifty years

Most of the reasons for these risks are self-explanatory. Those living in Denver are exposed to more background radiation than are residents of New York. There is natural radiation from stone and brick. Aflotoxins are present in peanut butter. Chloroform has been found in Miami drinking water. Saccharin is present in diet drinks.

It is not likely that such rational approaches to the evaluation of risk will have much effect on the reaction of the public to perceived risks. Nuclear energy will probably be thought of as a major threat because it is mysterious and unfamiliar. But who would be against peanut butter or vacations in Denver or living in a stone or brick house? Yet, if the calculations reflect even relative danger, it is no more risky to live for 150 years within twenty miles of a nuclear power plant than it is to live in a stone or brick house for two months. Regardless of the absolute accuracy of the numbers, the consideration of the relative importance of the risks is significant. It emphasizes the problems involved in trying to eliminate all the risk from our lives.

It is curious the values that are used in accepting risk. In assessing the possible effects of the radiation exposure to the population near the Three Mile Island nuclear facility, Joseph Califano, Secretary of HEW, testified on May 3, 1979, before the Senate Subcommittee on Energy Nuclear Proliferation and Federal Services: "If the total population dose at Three Mile Island was 3,500 person/rem—the current staff estimate or even double that amount—the traditional theory predicts approximately one additional cancer death. This compares with 325,000 cancer deaths that would normally be expected from all causes in the population of two million people residing within 50 miles of the Three Mile Island. On the other hand scientists who believe that the traditional theory underestimates the risk of low level radiation would predict up to ten additional cancer deaths for this population."

As Mr. Califano points out, while one or even ten deaths is a very low statistical increase, it is indeed significant to those who are the statistics.

Reaction of the public indicated that not a single death would be acceptable, even at the price of nuclear power. Yet it seems perfectly acceptable—in fact the public has demanded—that saccharin should remain available, even though predictions of harm have been made. And the entire country seems to accept nearly 100,000 deaths per year as the price paid for the enjoyment of smoking cigarettes.

Differences in reactions to different kinds of risks should be accepted. The most obvious classification of risks is that of voluntary risk and involuntary risk. Smokers now realize that when they smoke they are risking cancer. When they expose themselves to cancer they do so voluntarily. However, when workers are exposed to carcinogens in the workplace, or when consumers purchase products containing chemicals of doubtful safety, or when pollutants are loosed into the air to be breathed by everyone in the vicinity, the acceptance or rejection of the risk is no longer under the control of the individuals. They have no choice but to be exposed to the risk. Thus the individual who smokes three packs of cigarettes a day might be the same individual who protests vigorously the release into the air of any amount of radioactivity from a nuclear power plant. The ideal of making all risk-taking voluntary cannot, for obvious practical reasons, ever become a reality. The next best thing is to have the public as completely informed as possible about the degree of individual risks so that while not making individual decisions they can participate in decisions that do affect the public welfare. Such was the case with the acceptance of saccharin, and such could be the case with other products or situations now thought to be hazardous. But it requires more than simply making information available to the public. It requires that the public accept the responsibility for judging the consequences of accepting or rejecting the information as presented on an objective, not an emotional basis, a responsibility the public has shown little tendency to accept so far.

In order to make decisions based on risk-benefit analyses, in addition to the necessity to measure quantitatively what are the benefits versus the risks, we must face the reality of being able to answer a question that runs counter to our culture, our basic beliefs, even our humanness. What is the value in dollars of a human life? Or, to put the question even more coldly, how many dollars in savings should be exchanged for a human life?

When an economist with OSHA said that the agency's "overall goal should be a workplace free of carcinogenic risk" she meant just that—a totally risk-free environment. Totally risk free means that not a single life should be risked. When the courts reversed the OSHA regulation that required industry to reduce the exposure of workers to benzene in order to prevent an estimated one case of leukemia every six years at a cost of $300 million per case, it was deciding that here, at least, a life would not be worth $300 million.

A new emissions standard controlling the pollution from coke ovens in steel mills has been instituted. Proponents say that they will prevent about one hundred cancer and respiratory deaths a year. The cost to the producers will be about $240 million per year. The cost, then, per life saved would be $2.4 million.

Various groups, including the government and insurance companies, have attempted to calculate the value of a human life. For the most part they have used economic data to estimate the loss to an individual and society when a life is ended by accident. For example, the National Highway Traffic Safety Administration, using age and income data, estimates that each highway death costs society $287,175 in productivity loss, funeral expenses, and other costs associated with the accident.

Acrylonitrile, a product used in many plastics, causes cancer in rats. To eliminate the risk to factory workers OSHA decreed that air in the plastic factories contain no more than two parts per million of the chemical. The government calculated that, on the basis of data extrapolated from rats, about seven deaths per year would be saved, and that such a program would cost industry $24.3 million, or $3.5 million for each life. Similar calculations showed that the cost of reducing the level from 2.0 ppm to 1 ppm would be $29 million for each additional cancer death averted, and going from 1.0 ppm to 0.2 ppm would cost $169 million for each additional death averted.

Unfortunately our standards now do seem to be aimed at eliminating anything that anybody might think could cause harm. Unfortunately, too, the standards for determining the admission of products or processes into use are set to eliminate harm to the most susceptible individual in our society who might be exposed to them. We have, indeed, an example of protectionism run amok.

Since 1970 Congress has enacted eight laws applicable specifically to toxic substances. These are:

1. Occupational Safety and Health Act, 1970
2. Clean Air Act, 1970
3. Federal Water Pollution Control Act, 1972
4. Federal Environmental Pesticide Control Act, 1972
5. Federal Insecticide, Fungicide, and Rodenticide Act, 1972
6. Safe Drinking Water Act, 1974
7. Resource Conservation and Recovery Act, 1976
8. Toxic Substances Control Act, 1976

The task of assuring that every chemical produced in this country is free from risk, or even of discovery of what the risks are, is enormous. Consider as an example the Toxic Substances Control Act (TSCA). Enacted in late 1976, the law applies to chemicals not controlled by other laws. It requires that a manufacturer notify EPA of its intention to make a new chemical. Information on possible hazard is supplied by the manufacturer. Within ninety days the EPA must then decide whether the new chemical presents an "unreasonable risk to human health or the environment." It is an impossible assignment.

Consider the enormousness of the task. No one knows how many chemicals there are. The American Chemical Society's Chemical Abstract Service in November of 1977 listed 4,039,907 distinct products.[5] The number is growing at an average rate of 6,000 per week. EPA estimates that as many as 50,000 of these may be in everyday use, not including pesticides, pharmaceuticals, and food additives. There may be as many as 1,500 different active ingredients in pesticides. FDA estimates that 4,000 active and 2,000 inactive ingredients are used in drugs, 2,500 additives are used for nutritional value and flavoring, and 3,000 chemicals are used to promote product life. The best estimate is that there is a total of about 63,000 chemicals in common use. The first job of EPA is to obtain an inventory of products presently being manufactured. Unbelievably, there is no single source for this information now. Again, it is only possible to estimate the number of new chemicals being manufactured each year. These range from 100 to 1,000. But even 100 new chemicals represent a formidable task for an agency with the responsibility for finding whether or not a chemical presents risks. Simply answering the question "Does the chemical cause cancer in animals?" is a staggering amount of work. If the same requirements for proof of safety are to be applied to these products as are applied to drugs, —and logically they should be at least as stringent—then every chemical would have to be tested for the lifetime of mice and rats, and for seven years in dogs and monkeys.

Consider the practical aspects of the activity. The EPA must respond in ninety days. Assume that only five hundred applications are received each year. Then an evaluation of about two chemicals would have to be made every working day. EPA can respond to a request for approval in three ways. They can approve, they can reject, or they can request more safety data. I think the result is predictable. With the ninety day requirement the agency will be afraid to say yes, and there probably will not be enough obvious evidence to say no. Therefore, almost inevitably, more and more data will be requested.

There has been some question as to whether or not the making of new chemicals for experimental purposes of any kind might be interpreted as coming under this, or some other law. Every year there are thousands of new chemicals made in the laboratory for experimental purposes. For the most part their physiological action cannot be predicted accurately. In my own laboratory career I myself made over 500 chemicals that had never been made before and, for the most part, have never been made since. The same question concerning protecting laboratory workers was raised when trying to decide whether restrictions on toxic materials in laboratories should be enforced with the same rigidity as they are in manufacturing areas.

I have never considered it the responsibility of anyone but my co-workers to protect me while in the laboratory. In working with the unknown the possibility of harm is simply an occupational hazard that is, I believe, accepted by most scientists. The only way I could be totally protected is to be prevented from doing anything new, and that would be intolerable.

Do we dare even strive for a risk-free environment? I think even striving would have unfortunate consequences. It would mean no more advances by trial and error, since no trial could be made without a guarantee against error. No experiment could be performed unless the results were predictable. But the reason for performing an experiment is not to confirm something already known but to discover something unknown. The human race is not mean to progress or to survive without risk. Risk is a necessity for biological existence.

But striving for the elimination of all risk also carries with it a critical risk—the risk of stagnation, of over-dependence, of the stifling of originality and creativity, of discouraging change. To be sure of being risk-free means eliminating those things that cannot be guaranteed to be safe. But how can we be guaranteed that anything is safe? Scientists know that they can never guarantee against risk, particularly when risk is now defined as meaning any untoward effect that might happen to any individual under any circumstances. Consequently, when the question is asked: "Can you guarantee the safety of a nuclear energy plant?" or: "Can you guarantee the safety of a new drug?" or: "Can you guarantee the safety of a new automobile?" or: "Can you guarantee the safety of a television set?" the answer must be "No." Predictably when the answer is given the interpretation is that the plant or the drug or the automobile or the television set is unsafe. With all the wisdom of the ages scientists cannot guarantee that at no time under any circumstances will someone not be injured by any single product or process. When in their scientific honesty and political naiveté they say nothing can be guaranteed to be risk free they are astonished to find that they have become the source of the proof that everything is dangerous.

Assuming we could, at whatever cost, eliminate all risk, dare we introduce a generation into such a world? The world would say: "Do not worry, nothing can harm you, for we have guaranteed it. No harm can come to you for there is nothing harmful here. The only cost to you is that you will not be able to make decisions as to your own well-being. You will not be able to do anything that might have unpredictable results. You'll probably never have the inexpressible satisfaction of making a new discovery or doing anything that will truly advance the quality of life in the world.

"But you will be safe."

Human beings were not made to live like that. Risk taking is part of maturing. Even failing is part of maturing.

I don't believe the general public wants the kind of protection that is being attempted. Given the proper information, individuals are willing to accept the responsibility for taking risks, and are willing to accept the consequences of that risk taking. It is only when problems are made mysterious or are sensationalized that it becomes impossible to arrive at a rational decision as to the risk involved.

The general public still looks upon science and the scientist, regardless of what they perceive science and the scientist to be, as contributors to the common good and as solvers of major problems.

In a yearly poll taken by Yankelovich, Skelly, and White, favorable responses of between 81 and 84% were given every year since 1974 to the statement: "Science and advanced technology have brought us more benefits, through better products and an easier, healthier life, than the problems they may have created." In early 1980 an ABC News-Harris Poll found that 89% of Americans are convinced that "scientific research" will be a major factor in American greatness in the next twenty-five years. At the same time, the number of people believing that "Having a government which regulates industry" will be important was only 27 percent. In 1977, 36 percent of the public had thought government control would be important.

A survey commissioned by the Marsh and McLennan firm in 1980[6] gives a fascinating pattern of reactions to what the public believes science will be and do in the future. In spite of the apprehensions of increasing risk, the public does seem to be willing to accept that risk because they believe the results of science and technology will do more good than harm. Conducted by Louis Harris and Associates, the survey included 1488 members of the general public, 402 top level corporate officers, 104 institutional investors and bankers, 47 members of Congress, and 47 members of federal regulatory agencies. While 78% of the public think that society faces greater risk today than it did 20 years ago, only 38% of the business leaders feel that way. Only 24% of the business leaders expect "somewhat greater" risk from science and technology over the next two decades compared with 55 percent of the general public. However, in spite of the belief of greater risk, a majority of all classes surveyed believe that the benefits of science and technology outweigh the risks. Business leaders were overwhelmingly optimistic with 91% agreeing, as did 84% of the investors and bankers, 62% of the members of Congress, 68% of the regulators, and 58% of the public. While 56% of the public feel that technological development must be restrained to promote safety, 68% agree that technology should continue in "as uninhibited a regulatory environment as reasonably possible," and 53% feel that risks have been exaggerated by such events as Three Mile Island and Love Canal.

Probably the most revealing response of the survey is the attitude expressed toward the control of risk. When asked who should assume prime responsibility for ensuring a safe society in the future, 42% of the public agreed that the individual citizen should decide, while only 14% thought that business and 35% thought that government should be responsible. Even in such emotional issues as whether cancer-causing agents should be banned from food, only 10% of the general public were in favor of banning all these substances, while 46% would allow individuals to make their own decisions and 43% would decide whether or not to ban a product based on special circumstances.

When the first person used fire for benefit, no concept of either the benefit or the risk was contemplated. We now know the risk of using fire. When the wheel was invented, risks were not taken into account. But we now know that a single use of the wheel—the automobile—causes the deaths of 50,000 human beings a year in this country alone. The use of coal and oil for power, the conversion of coal and oil and radioactive substances into electricity, all carry tremendous risk. Even a drug that has been used safely on millions of people is still a risk. And the cigarette smoker knows that he or she might be one of the 100,000 people who will die in 1981 because of the cigarette.

There is nothing intrinsically good about risk itself. The benefit to individuals and to nations comes with the right to choose to take or not to take a risk. No risk is taken without some belief that a benefit will result if an action is completed and a bad effect averted. Very little progress is made without risk, either financial risk, or physical risk, or even psychological risk. So to eliminate risk almost by definition is to eliminate progress. And what happens to a generation, believing they are risk free, when the first risk appears as it inevitably must? Assessing risk and accepting risk in order to gain a greater benefit are the ways in which people mature.

No, a risk-free environment is not for the courageous, the creative, the ambitious. In a country such as ours taking risks is an imperative for the continuation and growth of the kind of civilization we have developed through time.

There are those who would go back to the never-were good old days when we were not exposed to risk as we see risk today. But we were not risk free. What kind of risk do you want to talk about? Financial risk? Tell the duPonts and the Dows and the Lillys and the Squibbs and the Searles that there was no financial risk in starting a company. They did not have the government regulations to contend with, but they risked their entire fortunes, although small by present-day standards, to develop products, markets, and business management techniques.

Personal risk? We didn't have DDT then, or drugs to cure other infectious diseases, but we did have millions of people dying from malaria around the world, and, in this country, a life expectancy of about 50 years because so many infants died within the first year after their birth.

Risk to the quality of life? We didn't have atomic power or electricity or automobiles or insecticides. But we didn't need these things, because we were using coal mined at the expense of the lives of miners, and an oil lamp was sufficient to light a house, and the cultural life was such that it was not necessary to have constant contact with friends and neighbors, and it was thought only natural to share the fruits of agricultural labor with worms and bugs.

I would not return to those days for anything in the world. They were not good days. They were days that were accepted because nothing better was possible. I could never understand why it was, and by some people still is, thought to be a more noble expression of our nature to be painted holding a pitchfork than it is to be painted while sitting on a tractor.

5 . . . wherein creativity is a necessary national resource, and it is found in diverse places, but it is not always nourished, and creativity is not simply innovation

The creativity of the United States, as reflected in its research and development, its production know-how, and its business methods, is a national resource. On its continued nuturing depends the future of the country and the well-being of our citizens. Creativity is the quality that allows individuals to prosper. It allows nations to compete and excel. It allows individuals to grow.

One of the practical results, as well as one of the primary objectives of a creative people, is the developing of products or services. Products or services are the medium by which domestic and international commerce is carried on, and they are the means by which the quality of life and the standard of living are increased.

Where do our products come from? What are the consequences of not developing products? Where will they come from if we do not allow those who are creative to be creative?

We can divide the sources of technical ideas and products into five different areas in the United States: the academic research laboratories in the university, the government laboratories, independent research institutes, the independent inventor, and private industry.

Traditionally, the academic laboratories have been the source of basic research. These laboratories were not equipped, nor did the researchers desire, to develop products. The term "pure research" is descriptive of this work, and also descriptive of what most academicians until very recently have wanted their work to be. It is work that is untainted by any stain of commercialism. In fact, many university people refrain from work if it can be suspected that any practical results will arise directly from their research. However, just as the term "basic research" implies, this work is the base on which much practical research is built, and from which products eventually result. A recent example is that of recombinant DNA. The research in the beginning was not directed to any practical result. However, with the developing of techniques to manipulate DNA also came useful applications of the techniques. Insulin has been produced using recombinant DNA processes, as have other medicinals such as endorphins, morphine-like analgesics, and urokinase, a human enzyme used for dissolving blood clots. An organism has been developed that will "eat oil," with the result that oil spills might be decontaminated. Of course, the organism could also eat oil that is not spilled.

When a chemist determines the structure, that is, the molecular composition of a complicated chemical, it is usually done to demonstrate scientific virtuosity and to bring some prestige to the scientist to establish his or her scientific reputation. The determination of the structure of penicillin in itself was of no practical value, although it was a very real scientific achievement. However, knowing this structure then allowed other chemists to vary the composition of the molecule, thus making antibiotics with different activities, and it even allowed for the addition of portions of the molecule—called precursors—to the medium in which the antibiotic was being grown in order to increase the production yield and make the process commercially feasible.

Chemists synthesizing chemical compounds in the university laboratory usually make the compounds to study the chemical reaction or simply to make a unique structure. They have no capability for studying the practical application of the chemical. In the past, pharmaceutical companies have taken these compounds and investigated their physiological activity. Insulin, liver extract, Benedryl, Merthiolate, some antibiotics, and many more drugs are products that originated in academic laboratories and were developed in industrial laboratories.

In the early days of research the university investigator was the elite of the research world. However, quality is no longer the exclusive property of the university professor, nor is basic research, as contrasted with applied research, considered the area of work reserved for the elite. It is recognized that basic research by itself does not contribute to an increase in the quality of life. By itself it produces no drugs for the relief of the ill, nor new energy sources for the provision of comfort or increased production, nor new computers, nor new communication methods, nor new space explorations. It is now recognized that the same quality of technological effort must be expended to make research results practical as was expended in uncovering the original basic data.

But there has been a distressing lack of cooperation between those doing basic research in the university and those using these results in practical applications. Some of this is caused by what I consider to be an outmoded attitude relating to the importance of basic research. There are still academicians who consider any contact with industry to be prostitution of their professional integrity. The majority do not. But there remains a gap between college professors and industrial researchers.

Neither do academic researchers cooperate with each other. I still remember an experience I had, not unusual, of supplying organic chemical compounds from our laboratory to a pharmacologist in a medical school for tests in a medical program, while accepting compounds from an organic chemist in the same institution for test in our pharmacology laboratory in the same field. It would have saved a lot of trouble if the chemist and pharmacologist, neighbors in the same institution, had taken the trouble to talk to each other. This illustrates another problem that exists in academic research life. Each worker believes that his or her future, the opportunity to advance through the academic ranks, to reach the Valhalla of tenure, depends upon his or her record of publications. Ideas are guarded jealously. Any cooperation with a colleague means sharing credit with that colleague. Instead of an exclusive publication he or she has a joint publication. All-important credit is diluted. So rather than cooperation within an institution there is really competition, sometimes to the detriment of the advance of science.

The belief of academic researchers that anything useful should be avoided has been a major factor in preventing true cooperation between universities and industry. The opinion prevailed that industrial research was inferior to academic research and that industrial scientists could not work objectively because they were controlled by their employers. There is today in major industrial laboratories the same quality and the same objectivity as prevails in universities. There is not the same freedom to work on any problem desired, but this does not indicate lack of objectivity.

Some of the reluctance on the part of university researchers to cooperate with industry can be partially justified. The objective of any industrial organization is to make money. When a company does research it tries, wherever possible, to protect results of that research by means of patents. Consequently, results of continuing research are closely guarded secrets until all the data necessary for the submission of a patent application have been obtained. When industry has cooperated with academicians or has sponsored research in an academic laboratory they have expected the same protection. This has resulted in some cases in a conflict between what the university researchers consider their academic freedom and what industry considers its right to protection. Nothing is more prized by anyone in academic life than academic freedom. In research this involves the right to publish whatever and whenever the investigator chooses. Consequently, any request to delay publication is looked upon as an infringement of this academic right. As a practical matter in a large majority of cases no delay is necessary since, if enough data are available for the publication of the scientific paper, they are also satisfactory for a patent application. Where delays are requested it is usually for no more than short periods of time. I know of no instance where the results of academic research were permanently delayed except in cases of classified government research, and in these situations the researcher is fully aware of the requirements before the research is started.

Industry has no objection to publication. In the first place publication of industry-sponsored research brings prestige to the sponsoring company. Secondly, the patent itself is a publication, so that the actual issuing of a patent, or in some foreign countries the application for a patent, makes the information public. Third, industry has no desire to assume credit for the academic work. The paper is published with only the names of the actual researchers as authors so that they get the total credit.

It is unfortunate that the two groups who can, with cooperation, originate a concept and make it practical, do not see the advantages of working with each other. It seems to me that, with enlightened research management in industry and recognition by the academician that he or she really does have peers in industry, such cooperation is possible. I can see no reason why universities can't continue to do basic research and industry continue to apply the results of that research. And I believe there can and should be cooperation on individual projects and still allow this freedom.

There has been a continuous fear that emphasis on basic research has been declining in recent years. The importance of basic research is recognized by all segments of the research community and, indeed, by the entire nation. But what has been the justification for basic research? It is justified on the basis of the fact that it is the foundation on which practical results are built, and from which products and processes result. If the pool of basic information dries up, then practical results will cease. If this is true, and I certainly believe it is, it seems peculiar that those doing basic research should resist cooperating with those who might some time use the work. It is time to admit that basic research is not useless research, and that the objective of doing it is not just to advance knowledge but to obtain knowledge that can be used.

Much excellent research comes from the various government laboratories, but they cannot be looked upon as a source of products. They are devoted to basic research and testing. The National Institutes of Health, for example, have done brilliant research. However, they are concerned with the study of particular disease categories, and the study of treatment of diseases. They do not develop products for treatment. Again, the work that they do is the basis for further practical results. It is obviously easier to find a cure for a disease if the cause can be specified.

On the other extreme a government laboratory such as the Bureau of Standards does almost exclusively practical research, but again the research does not create new products. It establishes standards that existing or future products must meet. An exception to the statement that government laboratories do not produce new products directly from their research might be found in the Department of Agriculture, where new agricultural advances have been made directly by them. They are examples, as is industry, of laboratories that have the ability and resources to carry a project from the concept stage, through testing, to an agricultural application.

There are many independent research institutes doing excellent research. These are of two kinds. The first is the nonprofit institute usually doing basic research in some specialized area, such as cancer, cell biology, reproduction, etc. The second is the organization that does contract research either for other companies or for individuals. The idea for the work does not originate within these groups. A project is brought to them because they have assembled groups of individuals in specialties whose various talents can be brought to bear on the solution of an assigned problem. Thus, while they are not the originators of basic ideas, they do contribute importantly to the stage intermediate between concept and production. They serve as test organizations for large and small companies. The range of tests, or "screens" as they are called, includes everything from medicinals, pesticides, and agricultural products to physical tests such as the determination of hardness of material or the strength of a plastic.

The independent laboratories make important contributions to the development of manufacturing processes, and to the developing of formulations that make practical the application or use of chemicals. The independent inventor is and has been the greatest source of original ideas for new products. The life and times of this group of people will be discussed later.

Finally, there are the industrial laboratories with their research and development groups having the specific objective of conceiving and developing products for commercialization.

In terms of social gains, improvement in the quality of life, raising the standard of living, and contributing to health and welfare, the results of research and development must stand high on the list of those things that have contributed to our way of life.

I have mentioned the contributions that have been made to medicine as a result of research done here. What happens if that research is no longer productive? Where do we get the medicines of the future? I would first like to eliminate the federal government as a possible source of new medicine. Those who suggest that the problem of finding cures for diseases could be solved by setting up government laboratories for that purpose are simply ignorant of what goes into the creation of a new drug. The government laboratories are ideally suited to doing fundamental or basic or pure research in all fields of science. They can even develop fundamental data on which a new drug is designed. They can find the mechanism by which a disease is caused, thus making it easier to find a prevention for that cause. But they are not organized, nor can they ever be organized, to do what industry does in research.

In no place but industry is there the one valuable ingredient that makes for success. That ingredient is competition. It is the most unappreciated, probably the most unknown, factor. The competition among research workers can only be described as fierce. That competition is equalled by the competition of management in the financial area. It has been responsible for research workers devoting their lives to the solution of their problems, and it has been responsible for stimulating management, at least in the past, to give more than generous support to their research groups. Management has realized that the company that could do the best research first was the company that would move ahead of its competition. So it was willing to take risks for success.

If drugs do not come from our industry or from our government where do they come from? If they are to come, they will come from foreign laboratories. The trend is already here. It is a humiliating admission to make. It is indeed sad to see a country with such a proud history of accomplishment now depending on others for its advances. We did have the finest research in the world. The same quality of people are still available to do research. We still have the best organizations in the world. The only thing that has changed is the policy that determines how free the scientists are to work on major problems. There has been no lack of lip service paid to the idea that technological innovation is important to our well-being, our economic progress, even our survival. Neither has there been any secret, at least in scientific circles, of the fact that we are losing our technological superiority. For the past decade various government committees have produced reports identifying problem areas. But nothing has happened.

Is it important to maintain our technological capability in the face of world competition? It would be an economic calamity if we did not. One of the important factors contributing to the decline of the value of the dollar has been the steadily increasing deficit the United States has been experiencing in our balance of trade. In 1967 the U.S. had a $4 billion trade surplus. In 1977 we had a $27.8 billion deficit, and in 1978 the deficit had increased to $39 billion. America's strongest exports have been in products requiring high technology—agricultural products, jet planes, computer know-how. The explanation given by the bureaucracy is that we are now importing large quantities of oil that is basically more expensive than it was a decade ago. This appears to be a good explanation on the surface, but it just doesn't hold water. Germany and Japan import more oil in relation to their GNP than does the United States. Yet they continue to have positive trade balances. The real reason for our deteriorating balance of trade is that our exports are decreasing. In 1960 the U.S. had 18% of the market of the world's exports. In 1977 we had only 12.7%.

In 1979 the merchandise trade deficit stood at $29.45 billion. To give some idea of the importance of technological products, in the face of the overall merchandise deficit, we exported almost ten billion dollars more in chemicals than we imported.

In addition, the rate of productivity change has been steadily decreasing. From 1955 to 1965 productivity increased at an annual rate of 3.1%. This decreased to 2.3% from 1965 to 1973. In 1978 the rate was barely positive, reaching 0.4% according to the Productivity Center of the Chamber of Commerce of the United States.

At the same time manufacturing reached a growth of 2.2% in 1976. But our chief competitors in world commerce easily outdistanced us. Canada reached 3.5%, both France and West Germany were at 5.8%, and Japan was at 8.9%.

How did the most technologically advanced country in the world arrive at this stage? There are two ways to increase productivity. One is by capital investment in modern plants and equipment. The other is by the development of innovative processes that allow for economies of production. I have already mentioned the $80 billion to $100 billion that has been drained from our industry to meet regulatory requirements. Very few of these dollars contributed anything to an increase in productivity. In fact, it is more than probable that since this was simply increased cost for the same production, it contributed in a major way to our decrease. With such a drain it is hard to see how many industries could afford additional billions to replace equipment to make their operation more efficient. The same argument holds for the lack of the development of innovative manufacturing processes. If economic demands lead administrators to make decisions inimical to research and development, obviously nothing new is going to result.

The National Science Foundation estimates that industry

employs about 385,000 scientists and engineers in research and development of all kinds, about 67% of the national 566,000.[2] Should 67% of the technological capability of the country be inhibited from performing at its top potential?

The crude oil situation is so recently with us that there is no need to point out in this instance the devastating results of a lack of technological planning. We are at the mercy of a group of countries who can extract from us almost anything they want until we are creative enough to come up with a replacement for oil or eliminate the need for oil.

Everyone knows about oil. But how about the other natural products upon which our industry and our economy depend?

The International Economy Policy Association lists twenty-seven minerals and metals that are imported in quantities that make us effectively dependent on foreign sources. We import essentially all of our mica, chromium, strontium, cobalt, tantalum, columbium, manganese, asbestos, aluminum, platinum, and the platinum metals, including iridium, osmium, palladium, rhodium, and ruthenium. We import over 75% of our titanium, bismuth, fluorine, tin, and mercury. We import over 50% of our potassium, silver, tungsten, zinc, gold, antimony, nickel, cadmium, and selenium. Each year we use over four billion tons of minerals. Only twenty years ago it was half that amount.

Even before the oil crisis it was easy to visualize what effect the shortage of oil would have on our daily lives. We know what it is, we can see it, we use it to fuel the engines of our automobiles, to heat our homes, to lubricate machinery. But who cares about chromium or cobalt or manganese?

We should care. Chromium is irreplaceable in stainless steels and in high-temperature-resistant super alloys, steels, and alloys that are used in jet engines, petrochemical and power plants, and other critical products. Without chromium and manganese there would be no steel industry.

The electronics industry is the largest user of gold after the jewelry industry. Silver, platinum, palladium, rhodium, and osmium are essential in electrical contacts and in the present high speed data processing and telecommunications equipment.

Iridium, one of the platinum metals, is the most corrosion-resistant element known, and is used in data processing equipment operating under rugged conditions.

Cobalt is used to make the steel used in strong permanent magnets, in high quality heat-resistant alloys and super steel for turbine blades of jet engines, in silicone-carbon tools, and in petroleum catalysts.

Unfamiliar products such as zirconium and hafnium are used in nuclear reactors, rhenium in high temperature alloys, iridium in thermo batteries for spacecraft, standby missile power supplies, and as a catalyst.

If there can be an OPEC—Organization of Petroleum Exporting Countries—there could also be an OMEC—Organization of Metal Exporting Countries. Zaire and Zambia are the principal suppliers of cobalt. Russia and Cuba supply most of the rest. South Africa alone supplies most of our chromite and ferrochromium, platinum and the platinum group metals, ferromanganese, and other minerals. Russia is the second supplier of chromium and the platinum group. Zaire, Zambia, Rhodesia, and South Africa produce a large portion of the world's supply of chromium, cobalt, antimony, copper, diamonds, germanium, gold, manganese, platinum metals, palladium, rhodium, ruthenium, osmium, iridium, uranium, and vanadium. A combine of these countries with Russia would give them control of at least 40% of the world's production of strategic materials, and a virtual monopoly on thirteen of the most strategic materials.

OPEC controls 52% of the world's oil. We import more than 52% of twenty-seven minerals or metals that are vital to our industry and to our economy. We have already had our warnings. Just prior to the invasion of Shaba Province in Zaire by Angola in 1978, Russia bought up much of the world's supply of cobalt. With the interruption of cobalt production in Zaire, the available cobalt was sold at $25 a pound, about four times the normal price. The price increased to $50 a pound on the spot market, the market where there is an immediately available supply of a product that can be bought without prior commitment. The flow of some strategic material has been interrupted, fortunately for only short periods, because of fighting in southern Africa.

We had not prepared for a situation that allowed us as a nation to be humiliated before the world by nations controlling our oil supply. Our way of life was changed, our economy suffered. What preparations are being made for the time when some other nation might shut off its supply of one of our vital minerals? Suppose we had no hard steel, no catalysts for our commercial processing. During World War II we developed a substitute for rubber when our entire import of that vital product was shut off. But we were lucky then. It happened, not as a result of planning, but because an industrial organization recognized the importance of an experiment performed by a priest-scientist in an academic laboratory, and had the vision to appreciate the practical implications of the discovery, and the resources to develop a new industry.

Must we always wait for a crisis? Perhaps we won't be so lucky next time. We have the technological ability to attack these problems. But these are long-range problems that require more risk than any industry is now willing to take.

To retain and increase our position of leadership in the world will require returning to our leadership in creativity—not just innovation, but creativity. Innovation is a much abused word. It has been used as a synonym for creativity, for imagination, for invention. Creative acts are innovative, but all innovation is not creative. Whenever something is done for the first time it is an innovation. When a company changes the form of an already existing product it is an innovation. The introduction of spray cans to produce aerosols of products was an innovation. The replacement of aerosols by manual pumps was an innovation. King-sized cigarettes and waterbeds were innovations, and were important commercially. But they are not the kind of creative changes that are necessary for a people to assume world leadership in commerce and industry, and they are not the kinds of innovations that result in better quality of life for a nation's individuals.

This is being written the day after I watched an exceptionally fine television program. To commemorate the 75th anniversary of the founding of their organization, one of the Big Three automobile manufacturers sponsored a program entitled "A Salute to American Imagination." At least twenty-five of the top stage and television artists in this country put on an unusually entertaining program. They did the things they did best, and the things most of them have been doing for a long time. If it were to be only a tribute to imagination, a recognition of the fact that there is such a thing as imagination in America, it was a success.

But I couldn't help thinking that a more appropriate recognition would have been, not just a salute to imagination but a salute to creativity, and a salute to those who have been creative. For imagination is not creativity, and innovation is neither imagination nor creativity.

One of the features of the program was an interview with one of the astronauts. The astronauts are true and authentic heroes, and should be recognized as such. The nation is in their debt. But neither imagination nor creativity is a characteristic of an astronaut. An astronaut is the epitome of discipline. And discipline is the enemy of creativity. The space program required imagination, but it was the imagination of a president who on May 25, 1961 said that as a national goal the country should "commit itself, in this decade, to landing a man on the moon and returning him safely." And the space program required creativity. But it was the creativity of those unnamed who developed the sophisticated hardware necessary to carry a payload and to monitor and control every action of the carrier and its contents.

There could be a salute to creativity in the automobile industry, even though much of what is done now is innovative rather than creative. The change in a body design is an innovation, not a creation. The introduction of a hatchback or a sunroof is an innovation. But the invention of the automatic transmission was a truly creative advance. But who remembers—who has ever known—the name of the inventor of the automatic transmission?

I discussed earlier the change in research and development strategies of industrial organizations, the change from long-range projects to those bringing immediate return, the substitution of the search for completely new products by changing or improving presently available products. This is innovation, but it is not creativity.

I even distinguish highly imaginative solutions to problems as being on a lower level than true creativity. For example, the production of a nipple for a baby nursing bottle requires that a small hole be made in rubber. For years this was done by means of a hot wire. With the advent of laser technology it was shown that a pulsed carbon dioxide laser could produce the hole easier, more economically, and with improved quality. It did require some imagination to recognize that a highly sophisticated technology could be applied to a very unsophisticated product.

How is creativity or innovation or imagination or inventiveness nurtured? It is difficult to extinguish the burning desire of individuals to create. If they have survived the normal educational procedure that tries to train creativity out and conformity in, if they have avoided the stultifying standards that say change is bad because it makes uncertain our comfortably predictable schedule, if they have arrived at a degree of maturity that allows them to be convinced that they are free to think, the truly creative individual will want to create. Creativity of the individual does not have to be nurtured. The creative fire will always be there. But the fire will produce no benefit.

What must be nurtured is the environment in which creativity may occur. For just as a fire, though still alive, may be banked so that it only smolders, so also the creative process can be thwarted by its environment. And just as the smoldering fire can be made to blaze by exposing it to the right atmosphere, so also can creative talent be released by the proper appreciation and encouragement.

The results of creative action do cause change. Some of the changes do disrupt a carefully programmed life. But the human spirit was not meant to deepen in wisdom and understanding by dedication to the status quo.

Totally aside from the pragmatic necessity for change—new products for our comfort, new industries to establish jobs so that citizens may provide for their own financial needs, a strong economic system to allow us to compete in world markets—the human spirit demands change for its growth and satisfaction. Human beings grow in spirit when they are challenged. Most changes resulting from major creative events do challenge us. Some result in risk.

When we celebrated the tenth anniversary of the landing on the moon, voices were raised questioning the benefits from that achievement. Would it not have been better to feed some of the hungry here on earth? But is it not also important to feed the human spirit?

Whether it is to advance our national economic goals, or gratify our intellectual curiosity, or satisfy the spiritual needs that demand challenge, or that physical need that demands comfort and pleasure, creativity is a necessity. Some progress can be made by accident. Even a blind pig finds an acorn once in a while. But for the benefit of our physical well-being and the satisfaction of the needs of the human spirit we'd better realize that encouraging creativity will give satisfaction not just to the unusual few, but will make more meaningful the life of every individual on earth.

6 . . . wherein the position of the United States as the world's leader in technology is imperiled, and the evidence is there for all to see

The wasting of a leader, the decline of an institution, the crumbling of an empire, are all sad and tragic happenings. When the leader and the institution and the empire compose the entire scientific and technological capability of a country the result can be not only tragic but a national calamity.

Since shortly after World War II the United States has been the acknowledged and proven leader in science and technology around the world. As reported by the National Science Board, the policy-making body of the National Science Foundation, in a study titled *Science Indicators*, the United States produced 82% of the world's major innovations in the later 1950s. In the mid 1960s, about 50% of the total research done in the world was done in the United States.[1]

That is not the way it is today. The same National Science Report mentioned above showed that not only did the percentage of developments in the United States decline but the quality of the innovations also seemed to decline. By the mid 1960s the percentage of major innovations had dropped from 82% to 55%. The study also evaluated the advances as to whether they were "radical breakthroughs" or "major technological advances." In the period from 1967 to 1973 the rate of development of radical breakthroughs declined nearly 50% over the period from 1955 to 1959, while the "major technological advances" doubled. And today about 20% of the world's research is done in the United States.

The indicators of our declining leadership have been available for all to see for some time. The first indication of the slackening of emphasis on research and development comes from the decreasing investment in research. For at least the last ten years, the percentage of our gross national product devoted to research has been falling—from about 2.9% in 1968 to about

2.0% in 1978. During this time corporate spending remained steady at about 1%. The government, once supporting about two-thirds of the total research, now supports only one-half. The proportion of scientists and engineers in the United States dropped from about 25.4 per 10,000 population to 24.8 per 10,000 in the period 1965 to 1975, while the proportions nearly doubled in both the Soviet Union and West Germany.

There are about ninety federal agencies with a total budget of $25 billion per year sponsoring research in this country.[2] About half of that Federal budget will be spent for work done in industry, so that about 35% of the total industrial R&D is funded by the government. The National Science Foundation in its publication *National Patterns of Research and Development Resources*

TABLE 1

National Expenditures for R&D 1960-1979
(Current Dollars in Millions)

Year	Total	Federal Government	Industry	Other
1960	$13,523	$ 8,738	$ 4,516	$ 269
1961	14,316	9,250	4,757	309
1962	15,394	9,911	5,123	360
1963	17,059	11,204	5,456	399
1964	18,854	12,536	5,888	430
1965	20,044	13,012	6,548	484
1966	21,846	13,969	7,328	549
1967	23,146	14,395	8,142	609
1968	24,604	14,926	9,005	673
1969	25,631	14,895	10,010	726
1970	25,905	14,668	10,439	738
1971	26,595	14,892	10,813	830
1972	28,413	15,755	11,698	960
1973	30,615	16,309	13,278	1028
1974	32,734	16,754	14,854	1126
1975	35,200	18,152	15,787	1261
1976	38,816	19,628	17,804	1384
1977	42,902	21,649	19,739	1514
1978	47,295	23,815	21,780	1700
1979	51,630	25,715	24,050	1865

Reference: *Science Indicators*, 1978, pg. 171.
Report of the National Science Board, 1979.

estimates that industry spent $36.7 billion on research and development in 1979, about 71 percent of the national total of $51.6 billion. About $24 billion of that will be industry's own money, the remaining $12 billion will come from sales of research and development to the government. An additional $280 million will go from industry to support work in universities and other nonprofit organizations.[3]

What has been the history of support of the various scientific sectors in this country? Table 1 shows the rounded-off figures from the National Science Foundation for support of the various sectors of the research community, as well as the amount of money spent by each. There is a constant and appreciable increase in expenditure over the years. The total research invest-

TABLE 2

National Expenditures for R&D 1960-1979
(Constant 1972 Dollars in Millions)

Year	Total	Federal Government	Industry	Other
1960	$19,693	$12,725	$ 6,576	$ 392
1961	20,664	13,351	6,866	447
1962	21,820	14,048	7,262	510
1963	23,829	15,651	7,621	557
1964	25,930	17,241	8,098	591
1965	26,970	17,508	8,811	651
1966	28,460	18,198	9,547	715
1967	29,291	18,217	10,303	771
1968	29,798	18,077	10,906	815
1969	29,556	17,176	11,543	837
1970	28,355	16,055	11,426	874
1971	27,697	15,509	11,261	927
1972	28,413	15,755	11,698	960
1973	28,937	15,415	12,550	972
1974	28,214	14,440	12,803	971
1975	27,684	14,276	12,416	992
1976	29,019	14,674	13,310	1035
1977	30,296	15,288	13,939	1070
1978	31,136	15,678	14,338	1119
1979	31,772	15,824	14,800	1148

Reference: *Science Indicators*, 1978, pg. 171.
Report of the National Science Board, 1979.

ment almost doubled between 1960 and 1970, and again between 1970 and 1979.

But things are not what they seem. If the expenditures are converted to cost of 1972 dollars a different picture emerges. From 1960 through 1969 total expenditures increased 55%. From 1970 through 1979 the increase was only 12%. In the last ten years the investment by the federal government actually decreased, as shown by Table 2.

An analysis of the relative emphasis placed on basic research, applied research, and development does nothing to encourage the belief that industry is working on new things. Data included in the National Science Board's Science Indicators are given in Table 3.

From 1970 through 1978 there was a total increase of only $11 million—or less than 2% in real dollars—spent for basic research by industry. In the same period applied research was increasing by over $500 million or 20%, while development increased by more than $2 billion or almost 29%. The inescapable conclusion is that not only is basic research still a very small percentage of total industry research, but that it is decreasing even more in importance compared to applied research and development.

Not only is basic research being eliminated in industry, it is also being cut back at universities, the traditional home for such work. The National Science Foundation reports that the $417 million spent in 1971 has declined to about $395 million in 1978, based on constant dollars. The result of this combination of spending is that industry has just about kept up with inflation, thus not expanding its efforts, while the source of much new basic information on which to build new invention has been curtailed.

In April of 1978 the problem of declining innovation was recognized at the highest levels of government when President Carter assigned to the Commerce Department the task of examining ways that the federal government could use to reverse the downward trend. The usual panel of business executives, public interest groups, and labor leaders was assembled and, after eighteen months, a plan that the President said would "revitalize America's industrial base" was presented. Other similar studies had been conducted in the past, but never at the high government level of the present domestic policy review.

TABLE 3

Industrial R&D Expenditures
(Dollars in Millions)

	Basic Research		Applied Research		Development	
Year	Current	Constant	Current	Constant	Current	Constant
1960	$342	$498	$1,226	$1,786	$ 2,948	$ 4,293
1961	361	521	1,195	1,725	3,201	4,621
1962	394	559	1,470	2,084	3,259	4,619
1963	425	594	1,483	2,071	3,548	4,956
1964	434	597	1,593	2,191	3,861	5,310
1965	461	620	1,654	2,226	4,433	5,965
1966	510	665	1,841	2,398	4,977	6,484
1967	492	623	1,889	2,391	5,761	7,290
1968	535	648	2,125	2,574	6,345	7,684
1969	540	622	2,320	2,675	7,150	8,245
1970	528	578	2,426	2,655	7,485	8,193
1971	547	570	2,492	2,595	7,774	8,096
1972	563	563	2,614	2,614	8,521	8,521
1973	605	572	2,889	2,731	9,784	9,248
1974	649	559	3,326	2,867	10,879	9,377
1975	688	541	3,519	2,768	11,580	9,107
1976	760	568	3,992	2,984	13,052	9,758
1977	840	593	4,412	3,117	14,485	10,229
1978	895	589	4,850	3,193	16,035	10,556

Reference: *Science Indicators*, 1978 pgs. 179, 180, 181.
Report of the National Science Board, 1979.

Nothing very startling was expected from the study. The expectations were realized. Important aspects influencing the decline of innovation were recognized, but the recommendations, such as they were, were little more than window dressing. Information centers would be established for the transfer of information, the patent system would be strengthened and uniform patent policies in government established, antitrust policies would be clarified but not reformed, the National Science Foundation's Small Business Innovation Program would be expanded, the federal procurement policy would allow the purchase of products from small firms, and so forth. Nothing in the President's plan referred to the economic issues that interested business most—reform of antitrust policy to make joint ventures easier, and restructuring tax policy to increase investment capital by giving tax breaks for research costs, and reduced capital gains on small business that made money on new inventions. About the best that can be said for the study is that it is at least a first positive step, that it did publicize the problem, and it did inferentially admit that the federal government did have some responsibility for the stimulation of innovation.

Who does the industrial research in the United States? The National Science Foundation and the Bureau of Census estimate that the number of companies doing some research and development exceeds 10,000. However, about 1,200 of these have 1,000 or more employees and account for about 95% of all the research and development work performed. Within these 1,200 companies, 100 companies account for 80%, the top eight companies take care of 35%.

It is sometimes dangerous to draw conclusions based only on the amount of financial support of an activity without being aware of what the money is being spent for. If total real dollars increase with the same quality of work being performed, then the results should be better. If total dollars increase with the quality being lowered, better results could still be obtained if the increase in dollars was sufficient to overcome the decrease in quality. If the support decreases, the only way an increase in results could be obtained is if there is an increase in quality more than sufficient to overcome the decrease in dollars.

I have, perhaps, shown my bias in using the above terminology—"decrease in quality"—when I mean going from research of new products to developmental research on old products. Certainly the quality in the real sense of scientific development can be just as high as that of scientific research. However, since we are concerned with the creation of new ideas, of new products, of advancing various scientific fields, I feel that qualitative judgment of the results of these two activities is justified.

To avoid the mistake of misinterpreting the significance of changes in research and development support it is necessary to know both who is doing the research and what is being done. For example, the defense industry and the space industry expend tremendous amounts of money for what is classed as research and development. Practically all this work is supported by the government. I would not class it as creative. The automobile companies spend a large part of the $3 billion credited to that industry's R&D budget for engineering design and styling, again a not very creative effort, and one that should doubtfully be called research.

Numbers do not indicate the quality of results. In the last four decades—the time within my own research career—occurred some of the most dramatic technological innovations in history. I can well remember in the early 1950s discussing with my neighbor, a young engineer, the project of his small company aimed at producing the first pocket radio. I owned one of the first such radios off the assembly line. It was made possible by the invention of the transistor, a product that has revolutionized electronics. The first commercial television is within the memory of most people living now. Computers are taken for granted. They are everywhere. In the same period such miracles as jet transportation, plastics designed to meet incredibly strict specifications, instant worldwide communications, instant photographs, lasers, and holograms, became part of our daily lives. Three factors were necessary for these developments. The first was the creativity and vision of the scientists and engineers involved, the second was the vision and willingness of those responsible for financing the projects to take risks. And they all required years of time from conception to reality.

There is ample evidence to support the conclusion that most companies are now drawing back from anything resembling long-range research—and I modify the definition of long-range to mean a minimum of as little as three years. We have the statements, almost universal, from company administration that they are committed to a yearly profit, growing at an appreciable rate per year. This can be accomplished by a close control of expenses, meaning that anything that doesn't pay off fast isn't done. Meaning that research isn't done.

Even a company as dependent on research as duPont has shown a "notable retrenchment" in its real dollar expenditure for research and development. A report on the company says: "In the process the company has shifted many of its research and development efforts from new venture research to work on established product lines."[4] The company has dropped about twenty-two new ventures, and is now working on only two or three. About 22% of duPont's research and development budget went for new product research last year, compared with 38% in 1972. Expenditures for "improvements for existing business" rose to 78% from 62%.

Raytheon reported in an interview: "Very definitely we have gotten away from long-term general research. All the research we are now doing is applied research—and a promise of payback within a reasonable period of time."[5]

In 1978 Zenith announced that it was laying off 25% of its work force, including a number of research workers. The research department was brought into the product development department. They announced the elimination of all research projects not directly related to their color television sets.[6]

At a meeting of the American Association for the Advancement of Science in 1979 Robert A. Charpie, President of Cabot Corporation, said that his firm had decided to put more emphasis on purchasing or licensing technology developed by others, rather than committing company funds to internal development. He said that uncertainty over markets and government regulations involved too great a risk.

Even the National Science Foundation, the body charged with the responsibility of furthering basic research in this country, has been accused of becoming too conservative. At a hearing in early 1979 before the House Subcommittee on Science, Research, and Technology, scientists testified that the Foundation seems now inclined to support "conservative proposals and proposals which are 'sure bets' in that they are most liable to provide some definable product in a short period of time," that the Foundation seems to be responding to pressure to "show that they have supported maximal numbers of programs which have paid off with evident successes." The pressure further caused the agency to decline to support "longer shots or more imaginative or longer term projects especially if a reviewer has given the proposal poor marks."

It sounds like a replay of what I said about business. Possibly I shouldn't be surprised. There is no reason to believe that the administrators of a scientific grant program should not be susceptible to the same kinds of pressures as the administrators of a business. In both cases short term successes are the measure of the organization's overall success. Both organizations have the same constituency pressures—the public and the government. The National Science Foundation is even more vulnerable to congressional pressure than is business, since it is responsible for its operation to Congress. Even objective scientists might tend to placate a congressman who gives out Golden Fleece Awards even if the basis for the award is to what extent the technical title of a research project impresses his technological ignorance.

Another indicator measured in the study of the Division of Science Resources was the patent balance—that is, the number of U.S. inventors patenting in foreign countries as compared with foreigners being granted U.S. patents. In 1978, 26,000 or about 37% of all U.S. patents were granted to foreign inventors. In 1961 the number was only 17%. The last ten years have seen a constant drop in the real number of U.S. patents issued to U.S. residents. In 1971 the number peaked at about 56,000. In 1978 it was about 41,000, while in 1979 it was down to a startling low of only 31,000.[7]

The decline of productivity of new products from just one industry, the pharmaceutical industry, will illustrate the several factors now converging to make it almost impossible under any practical circumstances to create or discover, develop, test, and market a new product. I use the term "decline of productivity" rather than "decline of creativity" advisedly. I believe people of great creative potential still do research in industry. However, for reasons I shall discuss, they just aren't being allowed to create. I use the pharmaceutical industry as an example because it has always been known as an industry that was built on research and creativity, that depended for its growth on creativity, and that required possibly more highly-trained scientists from more different disciplines than any other industry. Individual companies have spent as much as 11 or 12% of their total sales on research, and the average for all companies in the industry has been 8–9%. Compare that with the 1–2% that other industries, considered based on technology, have spent.

The late 1930s through the 50s are usually thought of as the golden age of pharmaceutical research. The sulpha drugs, antibiotics, antihistamines, non-addicting analgesics, polio and other vaccines, tranquilizers, cancer cures—all these miracles came out of the laboratories in a relatively short period of time.

However, even before this time, industry had shown its ability to recognize great scientific and medical potential, and then it had demonstrated its willingness to take risks to commercialize the discovery. Insulin and liver extract are examples. Both were discovered in academic laboratories. While still at a very basic level, the importance of the work was recognized by a pharmaceutical laboratory that then undertook the problem of converting the fundamental discovery into a practical reality. This was in the 1920s. Then in the late 1930s, came the sulpha drugs and very shortly after that penicillin. Then came the deluge of miracles.

What accounted for this special time in history, the period of unprecedented productive research? Many unrelated things.

First of all, pharmaceutical companies were small then. The depression had resulted in the failure of about 3,500 pharmaceutical firms. The 1,100 companies in existence in 1940 accounted for only $750 million in sales. In 1980 the industry sales were over $20 billion. It was not until the late 1940s that the first company reached the $100 million sales milestone. These companies were run by strong individuals. There were no research analysts to direct company policies indirectly. They were just as concerned as are today's leaders about their images. But they knew they were good because most of them had built their companies. So they were willing to take risks, even make mistakes. Thus, decisions were based on a more objective evaluation of the benefit of a particular activity to the company than are those made by present-day leaders.

The objective of most companies then was not just to make a continuing profit but to contribute something important to medicine. In spite of the fact that many of the leaders of companies were what might be called rugged individualists, there was still a certain amount of idealism present that helped direct their decisions. Thus, they were willing to take a chance on long-term, expensive research if there was the possibility that it would result in advancing medicine.

Research in medicine was starting from a low base. It is hard to remember the time when there were no antibiotics or sulphas or tranquilizers or antihistamines, when laboratories didn't have electron microscopes and CAT scanners and radio immune assays and dialysis. There was even a time when there were no computers. That's where we were in 1938.

Prior to World War II the United States was almost totally dependent on Europe for its research, including that on new medicines. However, in the early 1940s two things happened that resulted in the shift in leadership in the development of new drugs from Europe to the United States.

With the discovery of the sulpha drugs and penicillin a whole new atmosphere in drug research was created. For the first time researchers could have a hope of curing infectious diseases. This hope brought with it a new enthusiasm for the approach to the treatment and cure of all diseases. And, since the field was entirely new, it was an area where United States scientists could start off on an even basis with their counterparts around the world.

At about the same time, as a result of the war, the flow of research results from abroad on which American industry depended ceased. With the encouragement produced by the revolutionary breakthroughs in penicillin and the sulphas, combined with the necessity to do its own research for survival, the pharmaceutical industry began what was to be the development of the most creative and productive research program in the world.

The regulatory climate was different. There was a Food and Drug Administration, just as much interested in protecting the public as is the FDA now. But just as research was starting from a low base, so, also, were the regulatory agencies. They could regulate only what was known, or what they knew could be made known. Yet, in looking back at the results that were obtained in those comparatively primitive times I must conclude that the FDA did a good job with the resources it had at hand. In those days the FDA recognized not only its responsibility to be a policing force, but also its equally important responsibility to encourage the introduction of safe, effective drugs to the market.

I suppose if I had to attribute the success of that period to any one thing it would be to an atmosphere of freedom. Company administrators were free to run their own companies, scientists were free to work on research, even long-range research, the FDA was free to make objective, scientific, nonpolitical decisions.

What has changed?

Everything.

I have already mentioned the forces acting to influence the decisions of industrial leaders. The leaders themselves are no longer the leaders of the past. In their desire not just to be approved but, at all costs, to avoid disapproval, they have become almost anonymous. They must have a profitable company, they must have profits today and not a year from now, and they must have a minimum percentage growth every year. What is the result? The tragic result is that it is now impossible to think of long-term research on new, major problems in industry.

Regulatory problems have certainly contributed. Throughout the literature are recorded various estimates of the amount of money it takes to get a drug from the laboratory through the FDA's approval process. In October of 1977 Ronald W. Hansen, Assistant Professor of the Graduate School of Management, wrote the first definitive analysis of these costs for the Center for the Study of Drug Development at the University of Rochester. He obtained data from representative drug companies around the country, and calculated the time necessary for each phase, as well as the cost. Since the expenditures are spread over years, the costs were capitalized to the time of approval, using an 8% rate of interest. His figures show that, from the time of the first submission to the FDA until approval, the cost is $14 million in 1967 dollars, or $25 million in 1976 dollars. He further estimates that the expenditure is roughly the same to do the necessary work from the laboratory to first submission. His total estimate, then, is that the cost of developing a drug is now $31 million in 1967 dollars or $55 million in 1976 dollars.[8]

What has been the result of these changes? From 1940 through 1978 more than one thousand single entity drugs have been introduced in the United States market. Single entity drugs are used in this evaluation to distinguish what might more aptly be called medical advances from those products that are mixtures of already available material. Almost two-thirds of these, 64%, originated in the United States.

It sounds like a remarkable accomplishment, and it is. But a different picture might be predicted for the future if, instead of looking at an average of thirty-five years, we look at the trend of invention over these years. Table 4 indicates the years in which those single entity drugs were introduced.

The dramatic decrease in introduction starting in 1962 is apparent. The reasons for this decline are the subject of debate. The year 1962 was the year of the drug reform resulting from Senator Estes Kefauver's investigation. There can be no doubt that the increased requirements for drug testing both in time and money had a profound effect on drug introductions following 1962. However, even before 1962 there was some indication of a slowdown in real advances coming from the laboratories. It has been suggested that "all of the easy problems had been solved," that further advances would depend on new approaches. For whatever reason there can be no doubt that productivity as measured by the introduction of new products declined dramatically in the face of almost equally dramatic increases in research expenditures. In 1961 the industry spent $238 million for research.

TABLE 4

New Single Entity Drug Introductions to U.S. Market, 1940-1978

Year	Number of Entities	Year	Number of Entities
1940	14	1960	50
1941	17	1961	45
1942	13	1962	24
1943	10	1963	16
1944	13	1964	17
1945	13	1965	25
1946	19	1966	13
1947	26	1967	25
1948	29	1968	12
1949	38	1969	9
1950	32	1970	16
1951	38	1971	14
1952	40	1972	10
1953	53	1973	17
1954	42	1974	18
1955	36	1975	15
1956	48	1976	14
1957	52	1977	16
1958	47	1978	23
1959	65	Total	1024

Reference: *Fact Book: 1980,* pg. 30.
Pharmaceutical Manufacturers Association, Washington, D.C.

By 1979 this had risen to over $1.5 billion, a more than six-fold increase.

In addition to the single entity products introduced, numerous combinations of new and old drugs have come into use over the years. These, too, are declining in number. In 1958 there were 253 combination products introduced. There were 126 firms involved in the introduction of the single entity and combined drugs. By 1968 the combinations had dropped to 50, and there were only 48 firms who introduced a drug of any kind in that year.

Patents in the limited area of drugs and medicine might also be used as a measure of creativity. Table 5 shows the patents in this field issued in the United States, and the number of these patents resulting from work done in the United States and in foreign countries.

In 1963, two-thirds, 67% of all the drug patents issued, were the result of work done in the United States laboratories. By 1977 this had dropped to 54%. While the number of patents issued on United States work was more than double that of 1963,

TABLE 5

United States Drug Patents

	Originating in U.S.	Originating in Other Countries	Percentage of Total Originating in U.S.
1963	1034	498	67
1964	1180	622	65
1965	1182	683	63
1966	1703	849	67
1967	1637	801	67
1968	1164	500	70
1969	1654	976	63
1970	1596	941	63
1971	1509	908	62
1972	2292	1551	60
1973	1817	1349	57
1974	2059	1736	54
1975	2426	1959	55
1976	2448	2272	52
1977	2235	1933	54

Reference: *Fact Book, 1980*, pgs. 41, 42.
Pharmaceutical Manufacturers Association, Washington, D.C.

the number of patents from foreign work had increased almost four-fold. Thus, we have another indication that the research lead we had enjoyed here is rapidly being eroded.

Another alarming development that has taken place in recent years is the frequency with which drugs are introduced into foreign countries, sometimes years before they are made available here. Dr. William Wardell, Director of the University of Rochester Center for the Study of Drug Development, and Dr. Louis Lasagna, Chairman of the Department of Pharmacology and Toxicology of the University of Rochester School of Medicine and Dentistry, compared introductions in the United Kingdom and the United States for the period from January 1972 through June 1974. They selected nine major disease categories for which a total of 55 drugs had been made available. Of these, twelve were introduced in the United Kingdom first and seven in the United States. Twenty-six had been introduced only in the United Kingdom, and ten only in the United States.[9] Dr. Wardell continued this study and presented results through March of 1979 in testimony before the Science Research and Technology Subcommittee of the Committee on Science and Technology, U.S. House of Representatives, on June 19, 1979. The study showed that about four times as many drugs are being introduced exclusively into Britain as into the United States. Thus, even though drugs are available they are not being introduced here.

The cost of such late introduction could be enormous. An attempt has been made to calculate the loss caused by the delay in some specific cases. Professor Sam Peltzman, of UCLA, estimated that a two-year delay in having introduced the presently available anti-tubercular drugs would have cost about 13,000 lives and $2 billion.

Finally, the increasing trend toward doing research abroad should be cause for concern. In 1963 United States drug firms spent 93% of their total budget in the United States and 7%, or $18.9 million, in foreign countries. By 1971, the percentage was still only 8.3% or $52.3 million spent abroad. Between 1971 and 1979 the percentage figure more than doubled—to 18.0%, while the dollar figure increased more than five-fold to $269.9 million.[10] The National Science Foundation's Industry Studies Group, Division of Scientific Resources Studies, reports that the expenditure of all research being done by United States companies abroad increased 41% between 1974 and 1977, while research done in the United States by the same group increased only 32%. Research in the chemical industry grew 81% abroad in the same period, while pharmaceutical research rose 105%.

The Center for the Study of Drug Development at the University of Rochester studied the pattern of the investigation of New Chemical Entities (NCE) that had shown enough pharmacological activity to have reached the stage of clinical investigation. Between 1963 and 1969 only 8% of the United States NCEs were first tested abroad. In 1974 the number rose to 34% and it is estimated that it was 47% in 1975.[11]

In June of 1979 the General Accounting Office testified before a subcommittee of the House Committee on Science and Technology. They reported that, based on 132 new drug applications submitted in 1975, the average time involved for approval after clinical trial had been completed and an application submitted was 20 months. They also stated that of 14 "therapeutically important" new drugs approved by the FDA between July 1975 and February 1978 "thirteen of these drugs were available elsewhere two months to 12 years earlier than they were available in the United States."

There can be little doubt that both the introduction of new drugs abroad before they are available in the United States and the increasing expenditure of research dollars abroad are the direct result of restrictive regulatory policies. There is obviously merit in spending some research money abroad. The United States has no monopoly on scientific brains, so some money should go to support foreign brains. If a company is to operate in a foreign country it should become part of that country. Sponsoring research and development is one way of doing that. But I think we should be suspicious of a situation that encourages the spending abroad of almost $1 in every $5 of the total spent for human pharmaceutical research, particularly when the organizations spending the money have had such a magnificent history of accomplishment in past years without such an effort.

Every company in the pharmaceutical industry can look at its research program today and tell exactly the maximum number of new products it will introduce in the next eight to ten years. I say "maximum number" because probably a majority of those that have shown positive results in research will never pass the stringent tests required for marketing. If an idea has not already been conceived and reduced to practice, if a product is not already being tested in research today, it will not be available as a drug in the next eight to ten years.

It is because of these factors that I have said that there will be few major new drug developments originating in industrial pharmaceutical laboratories in the future as long as these conditions exist, or unless the research on that drug is already well under way.

I have said many times that the job of a research director is not to create but to select creative scientists and give them the proper conditions under which they can create. The same creativity is present in today's scientists as was present in the early days of the medical revolution. The atmosphere, alas, is so totally changed that creativity never has a chance to emerge.

If the situation is as I have described it, why is it that companies continue to increase their investment in research and development? The 23rd Annual McGraw-Hill Economics Department Survey of Business Plans for Research and Development, published in 1978, predicts that expenditures for research and development in the United States will continue to increase through 1981 as they have over the past years. For example, the industry producing chemicals, the industry generally conceded to require the highest concentration of scientists, spent $3 billion in 1976. In 1977 the estimated spending was $3.8 billion, while the same companies plan to spend $4.2 billion in 1978. Projections indicate that the investment in research and development will reach $5.23 billion in 1981. This group of companies includes those making industrial chemicals, drugs and medicines, and "other chemicals."

For example, why has the pharmaceutical industry, one of the most creative industrial research sources, continued to increase its research expenditures every year, with the last five years increasing from $973 million in 1975, $1.1 billion in 1976, $1.2 billion in 1977, $1.3 billion in 1978, and $1.5 billion in 1979? Why do they continue to spend 11.5% of their sales dollars on research and development?

The reasons are many and complex, logical and illogical. First of all, a major company will have had a research program to which it has been committed for a number of years. It is probable that there are problems that had been attacked for eight or ten or more years. Possibly there are just enough encouraging results to stimulate management to agree to continuing for "just another year," even though the "just another year" decision is one that has been made each year for some years before. There has already been a significant investment in the project, and the solution might be just around the corner. It is easy to start a research program, but it is one of the most difficult jobs in the world to discontinue it. Thus, some research is sponsored simply because of the past momentum developed by the program and the present inertia of management.

A second reason is that most industrial groups are organized to work in a particular area, and scientists working on a special problem are themselves specialists. It is not as easy as one might think to transfer scientists from one problem to another. They are expert because they have spent a great deal of time learning all there is to know about the field in which they are working. Therefore, the discontinuation of a problem will disrupt the working of a highly-trained group of people who possibly, then, cannot apply their talents to other interests of the company.

Third, there is a legitimate need to improve present products, to protect against competition, to do "defensive research."

Fourth, the cost of doing research has increased from year to year, both because of increased salaries, equipment, supplies, and everything else necessary to carry on a business, and because of regulatory requirements in some industries that make it necessary to do more to get the same results.

Finally, there is the wholly subjective judgment of management that has to do with company image, the way a company is perceived by the outside community. I believe most companies are afraid to cut back on research. A mystique has been built up around research. The companies that do the most research are the "best" companies. As long as research is supported, the company will be successful. If a company continues to increase its research budget year after year it must be successful.

I believe this was true in the past. When individual researchers were free to think, it is reasonable to suppose that, of two competing companies each with creative scientists, the one with the most scientists would produce more new ideas and more new products. Since the number of scientists was seldom published, the judgment on the quantitative factor of research was based on the published research budget. Thus, the company with the higher budget was judged to have a better product potential. It was then as it is now to the advantage of a company to publish as high a research budget as possible to impress the financial analysts with its dedication to science.

And so it is that managements have been reluctant to cut back on research. It is not for any logical business reason. It is because it might be interpreted to mean that the company is no longer interested in the future, that they will not be able to compete with other companies still increasing research expenditures. As a matter of fact they are not interested in the future. They are interested in the present. And an immediate present effect of an announcement of a research cutback would be that the price of the company's stock would decrease.

In many high technology industries there is no way that high research budgets can be justified on a business basis as long as the research is predominantly defensive research. In the past, long-range research was justified on the basis that major products would result and that these products would be sold for many years into the future. Thus, after marketing, the profits accumulated over years would pay for the research done in the past. For example, research done in year one would be justified on profits from products sold in years five, six, seven, and so on. Planning was done not on the basis of profits in the year the research was done but at some time in the future.

Now, however, research done in year one might have to be justified by the profits projected from sales in years ten, eleven, twelve, and so on. If we add to that the fact that most research is not directed to finding major new breakthroughs, and therefore the profit potential will be less, it is difficult to rationalize the continued expenditure of large amounts for research on any business basis.

I make a distinction between simply improving a product, or improving on the way a product is produced, and extending the technology in a product field. There is the same danger in neglecting technology extension as there is in neglecting new product research. And there is the same creativity required for new technology as there is for new products.

A recent and costly example of this neglect is in TV technology. For about the first fifteen years the United States dominated this field. For whatever reason, the Japanese have taken the lead in what promises to be a major new television development—the production of the video cassette that enables the user to record programs for later viewing.

In 1970 I was privileged to be one of a group of ten scientists asked to visit India to advise on the industrial research taking place in that country. The commission was arranged by the National Academy of Sciences and our State Department, at the request of the Indian government. We visited large and small research organizations in industry, and were welcomed as well in the government laboratories. It was a tremendously stimulating experience. It was also a very humbling one, not because of what I saw in the research, but what I saw in production.

From our sheltered and sometimes parochial position in the United States it is easy to become imbued with the feeling that we are the leaders in all fields of technology. It is something of a shock to find that other nations do not consider this true. During our travels we usually stayed overnight at some company guest house. Invariably, we were only a small part of a larger group also staying there. The other guests were Japanese, who were in India building steel and fertilizer plants for the government. The plants were to be built, operated for a period of time, and then turned over to the Indians for operation after all the difficulties had been ironed out. It was striking to me that, having the world to choose from, the Indian government did not select United States technology but Japanese technology for their country. And I could not help wondering, having seen some of our own World War I installations still being operated, how long we could continue to compete.

It does not do much good simply to recognize advanced technology. To change some of our plants into facilities that would have the same efficiency as the plants that I saw would require hundreds of millions of dollars.

The last large steel mill was built in this country at Burns Harbor, Indiana by Bethlehem Steel in 1960. U.S. Steel studied for five years the possibility of building a new plant in Conneaut, Ohio. However, in view of a staggering cost of between $3 and $4 billion, the company, in 1980, decided it could not make such an investment. In times of high steel demand even the ancient plants can be profitable. But when times of real competition set in they will be in trouble. What will happen in Youngstown, Ohio to the plant employing 3,500 people that still drives one of its rolling mills with a steam engine installed in 1908? With any weakening of an economy it will be marginal plants that will have to close. And the marginal plants are usually not those employing the best and most modern technology.

An example of maturity being a disadvantage to further progress is seen in the relatively new but recently explosive semiconductor industry. Since only about 1970 progress in this industry has resulted in revolutions in the development of calculators, analytical instruments, weapons, and space developments. To accomplish these things the industry found it necessary to invest in plants, equipment, and research and development. With such a heavy investment it is natural that they should be unwilling or even unable to change to procedures that would require even heavier investments without having had a chance to recover their original money. In the meantime, Japan, without having had to spend money for the initial pioneering work, can enter a developed market at a more advanced technology base and thus have more flexibility to take advantage of even further technical developments. The result will be that the United States industry will remain with the technology available now, while the Japanese will be competing with superior products. It is inevitable that the market will eventually go to the Japanese, and we will be forced out of the market that we have pioneered and developed.

It is not that our industry is not capable of the same innovative efforts as the Japanese. It is just that economic factors lead industry to arrive at the decision that it is not economically advantageous in the short run to reduce present profits in exchange for continuing in the market some time in the future.

If we are to consider research and development to be a national resource used not only to provide employment for the citizens but to be a major factor in allowing the country to compete with the rest of the world on an economic basis, it is not sufficient to know that we spend more money on research than any other country, nor is it of significance in itself that we have more engineers and scientists than any other country. We do have about 566,000 scientists and engineers. Industry uses about 385,000 or 67% of the total.[12] If we assume that each scientist and engineer around the world has the same production capacity, and therefore that the country with the most scientists should be the most productive, we seem to be in pretty good shape. However, if we compare creative and production potentials of scientists around the world we don't look so good.

Of the 385,000 scientists and engineers in industry about 120,000 or 31% are paid from government funds, indicating that they are probably on space- or defense-related projects. The technical effort remaining, therefore, for work that will contribute to the economy by the development of products for sale here and abroad is drastically reduced.

Japan is a serious competitor in technology and in economics. The United States has about twice as many scientists and engineers working in research and development as has Japan. But Japan spends less than 10% of its research and development budget on defense and space, while we spend over 60% of our government budget in the same areas. In 1975 Japan had only 25% fewer people doing research and development not related to space and defense than we had in the United States. When it is considered that Japan has a sharply-defined program aimed at advancing its economic interest, it should not be surprising that we now have a huge negative balance of trade with Japan, and that much of Japan's success has come about as a result of commercializing and improving on work originally done here. If scientists in U.S. industry are not used efficiently, the consequences to the future of the country could be disastrous.

Of importance also is the lack of initiative on the part of U.S. companies or the government to take advantage of the huge potential for the development of products based on research performed around the world. We have seen how effectively foreign companies use the results of our research. It is not by accident that technical results are available and are evaluated in foreign countries almost as rapidly as they are in the country of origin. The result for them has been so positive that it has resulted in complaints from domestic companies, and the suggestion that trade barriers be instituted to prevent the import of goods based on research and development results from our own laboratories.

The efforts of foreign companies not only to gather but to use our research results are systematic and effective. There is nothing either unethical or illegal about such operations. They do not involve industrial espionage. Practically every foreign government has attached to its embassy here one or more scientific liaison people whose responsibility it is to keep abreast of scientific progress and to report it to their governments who in turn inform industry, who in turn use it to their advantage.

France has six science attachés in Washington, as well as others in major cities in the United States. Industry in Japan is organized in such a manner that they can have technically-trained people in five of the major cities here. Individual Japanese companies carry on the same type of surveillance.

The United States does maintain scientific attachés at our foreign embassies. However, they are not involved with collecting useful information. I don't know what they do.

It would probably be considered a violation of anti-trust laws if U.S. companies were to cooperate in the collecting of information from abroad. A few individual companies have their own information-gathering system, but the majority either consider it not cost effective or are so self-centered that they do not consider that anything useful could come of it. Considering that well over half of the scientific developments now occur abroad, it is, indeed, a shortsighted attitude.

Finally, United States industry is at a disadvantage because of the indirect support some foreign industries receive from their governments. In 1973 West Germany, through its Ministry for Science and Technology, provided about $4 billion for product and process development. As long as the industry contributed at least 50% of its cost, it retained full rights to the results. If the project were profitable, some payment to the government was expected. Small firms retained full rights regardless of their contributions. For other projects the Economic Ministry provides interest-free loans up to 50% of the cost of technological innovation.

The Japanese government has selected certain areas of technology that they have decided are of national interest. Direct support is given for this work and, in addition, special low interest loans are provided by the Japanese Development Bank.

There now seems to be some recognition of the need for government encouragement at the basic research level in the United States. The National Science Foundation, the government body responsible for fostering basic research in this country, has taken the unprecedented step of supporting grants to industry. In the past, $850 million controlled by the Foundation has gone almost exclusively to academic research people for the study of fundamental research problems for which there is no objective of a commercial profit. In 1976 industry spent $786 million of its own money for just such basic research. This amounts to 3.5% of all industrial research and development spending, down drastically from 7% spent in 1963.

The need for basic research is recognized by everyone in the research field. It is the foundation on which later practical research is built. There are many examples of what appear to be useless research resulting in the conversion of a research concept into a practical product. When it was discovered that animal tissues could be grown in the test tube it was a fascinating research result. The company that first studied this technique in industry as a pure research tool became the company that was able to manufacture the Salk polio vaccine, using that process on a manufacturing scale. If, then, the level of basic research decreases it can be predicted accurately that the number of real advances will also decrease.

The decision of the National Science Foundation, therefore, to sponsor research in industry is a further recognition of the importance of this activity. There is one qualification added to this grant approval, and that is that the industrial work must be done in cooperation with an academic institution. The amount of money assigned to the project, $4.5 million, will not fund many research problems. However, it will be enough to demonstrate the advantages and disadvantages of the system.

I am convinced that the direct support of research in an industry by the government will not solve our problem. Under a direct subsidy arrangement the government would be forced to take control of the results, and to allow others to compete with the company that first developed them. Such a situation eliminates one of the great reasons for the success of industrial research, namely, its competitiveness. Having worked in large and small organizations I can assure you that the competition among research people is intense. A great part of the satisfaction that comes from research comes not just from doing something first but in doing something that your competitor can't do, in knowing something your competitor doesn't know, and in being able to use something your competitor can't use. The satisfaction derives from the fact that this has been a unique contribution from your laboratory, and that you are ahead of competition. Further, you deserve to be able to stay ahead because you have done something creative. Elimination of that drive would surely be a detriment to progress.

Government sponsorship of research in industry will not solve the problem as long as management is obsessed with the desire for short-term profits and as long as regulatory agencies overregulate and as long as the public accepts this overregulation in the desire for a risk-free society. If overregulation is an important factor in inhibiting risk-taking in industry, and I am convinced that it is, consider how illogical it is to try to solve the problem by government support of research. Research dollars will be spent in order to correct a situation that should not exist in the first place. The more money spent for regulation the more that must be spent to do research under those circumstances. Thus, the government is wasting money both in regulating and in doing research to correct the regulating. It is probably too logical to suggest that a more moderate view of regulation be taken, while allowing industry to spend its own money on what it formerly did best, that is, turning out more and better products then anyone else in the world.

7 . . . wherein venture capitalists do not want to venture, and entrepreneurs have difficulty surviving, and they are very important

There are, in business jargon, many expressive terms used to describe various activities. There is none that is so misleading as the term "risk capital." Practically every major organization has its Venture Capital division or subsidiary. Dozens of independent organizations are set up with the announced purpose of supplying "risk capital." Television programs extoll the wisdom of industrial organizations with the foresight to recognize frontier types of technological activities, and the courage to risk their money to develop new scientific breakthroughs that will eventually bring to everyone in the world a happy, healthy, prosperous life.

Don't believe it. There is no such thing as conscious risk-taking in the venture or risk community. At the first hint of doubt as to whether a project will succeed, venture capitalists run like mad, taking their money with them. The criteria used by a venture capital group should be the tip-off. First, they usually demand control over whatever organization is established to carry out the venture. There is nothing too much wrong with that as a concept. Whoever has the money is in the position to demand control. Why is control demanded? Not just to assure a profit, but to guarantee against loss.

In the ventures that I am talking about, those that depend upon a successful conclusion of a technical program, there are two different kinds of direction necessary. First is the direction necessary to solve the scientific problems and to produce a marketable product or service. The second is the concurrent problem of designing a plan to sell the product successfully in the marketplace.

If inventors are lucky enough to interest a financier in putting up the money to develop their brain children, they will all too frequently find that the concept of control means not only control of the business plan but also control of the technical plan. Many times this is not a conscious desire to control technical activities, but is a direct result of a control of the approval of money for the experimentation necessary.

When I speak of risk capital or venture capital I mean money to allow for the start-up of a totally new company, or money to provide operating capital for a company already started but requiring additional resources to survive. The obstacles facing the entrepreneurs starting their own companies are immediately apparent. They have no track record. Therefore, all judgments are based on what the products might be and what the entrepreneurs might do. If the entrepreneurs are presenting only an idea or a concept they have practically no chance for support, even though they have done considerable work to support the concept. Each will be asked, at the very least, for a demonstration of the working model, a prototype of the finished product. If the inventors are typical, they will not have approached anyone for outside support until they have exhausted their personal resources. Therefore, the request to do anything that would require more investment from the inventor is tantamount to a rejection of the project. The most critical need is for support to prove whether or not an idea or a project is good. If the venture capital source requires a model for support, the impasse is obvious. The inventor needs money to make the model, the venture capitalist needs a model to approve support.

But even proven prototypes, production experience, and field trials do not make the situation much easier. I can relate only a few of my own experiences to illustrate the difficulties. In considering these experiences the state of the development of my products should be kept in mind. It will give a much better appreciation for the lack of imagination used and the fear of any risk on the part of not only venture capitalists but already large operating companies.

1. The products had been developed by people with years of experience in the fields in which the products were to be used. My own experience alone covered thirty-five years.
2. No concepts were presented. All products were at least to the prototype stage, all had had production development completed, most had had clinical trial, and some had gone through semi-production.
3. Most had patents already issued on them, others had patents applied for.

4. All products were developed to fill a need.
5. Sales estimates were presented for each product.
6. All were either unique or had obvious, demonstrable advantages over an available product.
7. All could compete on an economic basis.
8. Both sales and profit are high, the sales potential in the medical field alone being over $200 million.

CASE 1. One of the non-medical products developed through two prototype stages required a development effort that was far beyond the reach of our resources. It was a project of immediate national and international importance, and required a familiarity with developing large rafts or islands located in the sea. Accordingly, I selected one of the five largest companies in the world as a first candidate for a possible collaborator.

I was acquainted with the president of the laboratories through activities in Washington, so I wrote to him detailing the facts that favorable action had been received on patent applications, and that small prototypes had been tested successfully. About six weeks later, with no intervening communication, I received a phone call from someone in New York who identified himself as being in charge of the Venture Capital Division of the company. This sounded good, since a company of that size should be able to devote the proper resources, and have available the proper talent, to analyze the project I was proposing. I made an appointment to meet with their expert.

I arrived at his office promptly at twelve noon on the hottest day in the summer of 1977. After exchanging the usual amenities my host suggested that we walk to the restaurant for lunch, not far away. He also suggested that I take my briefcase, crammed with all the documents I could carry, so that we could talk through lunch.

We started walking. After about a half hour it was apparent that my guide did not know where we were going. Finally, the combination of the heat and the weight of my bag dictated that I should drop all pretense of dependence on my guide. When I asked him where we were going he said "to the Four Seasons." I directed him to the restaurant. It is one of the most expensive eating houses in New York and is a kind of place you go to either impress someone who has never been there before or else to take advantage of a situation that allows the host to go there for the first time. I had already been there.

During lunch I explained the workings of my process. With

no data available to him, the reaction of my host was that the process was too expensive. I explained that a commercial device would cost roughly $300,000, and would serve the same purpose as a plant that had just been contracted for at a cost of $32 million in one of the Near East countries. The next reaction was that the process wouldn't work. I explained that two prototypes were already working in two different parts of the world. By that time lunch was almost over, and I suggested that possibly this project didn't fit into their policy of support. I confess now that maybe what I really did say was that their policy didn't cover projects involving new and creative technology. He assured me that this was not so, and gave as an example the fact that they had just invested in a coal mine that had been thought to have been exhausted.

In comparing backgrounds I found that this was the first position my host had ever had, and that he had received his Bachelors Degree in Mechanical Engineering from a technical school—not MIT—only two years before.

Following lunch, and with no more detailed discussion, I bade farewell to my host, hoisted my briefcase, still unopened, and headed into the sunset, this time in the welcome atmosphere of an air-conditioned cab. I assume my host found his way back to his office by himself.

CASE 2. One of the three largest pharmaceutical houses in this country had recently changed its policy that involved selling only drugs, and had acquired a company that made a hospital instrument. Believing that it was impractical to succeed with only one product in a new line, I contacted the company to see if they would be interested in taking over any of my products. I was well acquainted with their executive vice-president, and he agreed to meet with me to look over our developments. He flew to Chicago on his company plane, arriving in my office about 10:30 in the morning. He immediately announced that he could spend one half hour with me to discuss, see demonstrated, inspect data on, and analyze the twenty products I had at that time. After about fifteen minutes of frantic talking, we were interrupted by a phone call from the pilot who said it would take another two hours to get the plane in shape. Somewhat relieved, I started a more leisurely presentation. This went on for another half hour, when it became apparent that there wasn't the slightest interest in anything technical. The rest of the time was spent talking about families and mutual friends.

I received a prompt courtesy letter telling me that someone from the new division would be in touch with me within two weeks. That was two years ago. If I don't hear soon I'm going to stop waiting.

CASE 3. Most of the great family fortunes in this country have, as part of their diverse holdings, a company devoted to investing in new ventures. I was referred to one of these as a possible source of support for one of my products.

I was given an extremely courteous and friendly reception. When we got down to talking about what I required for my operation the first questions asked were: "What product do you manufacture?" and "What are your sales?" When I responded that the investment was to allow me to get into production and develop sales, it was explained that the policy of the organization was not to invest risk capital in anything involving a risk, but that support was given only to already operating companies whose potential was readily assessable.

CASE 4. I was interested in a general cooperation with a U.S. company owned by a foreign corporation. Within the corporate structure were companies in practically all areas of the health care field, and having representation in most countries of the world. Because we were also negotiating with several other companies on individual items I asked for an expedited decision from them. I made a full-scale presentation of all our products, and requested an expression of interest within two weeks. The answer was that they were interested, and would like more extended discussion and a chance for further analysis of the products. Samples were supplied, and all data were submitted.

After nine months of procrastination I was given the decision that they were definitely interested in eleven of the major products, representing about $183 million in potential sales. They made the proposal to me that I give them an option for six months to arrive at a decision during which time I would not have any discussion with any other companies. The option would include a worldwide, exclusive right to make, use, and sell the products.

For this, they would not even agree to an option payment.

CASE 5. In one of my weak moments I agreed to allow an organization to seek equity investment in my company. The first two investment groups approached gave the same negative reaction for the same reason—I had too many products and, therefore, could not concentrate on promoting them. This decision was strictly according to the textbook, and probably would have been applicable to a situation where I was marketing the product to thousands of customers rather than demonstrating developed products to a selected group of companies.

There were no more attempts to obtain equity capital.

CASE 6. Many of our products were developed using not only the highest technical competence but also very ingenious methods of assembly. After inspecting one handmade prototype, we were told by members of a company's technical group that it would be impossible to manufacture. When products were produced from a first soft mold we were told the product could not be produced economically. When quantities of samples were supplied from a semi-production mold and firm cost estimates were presented, the research department of the major organization involved still insisted it was impossible.

This is a modification of the NIH (not invented here) syndrome. This one says "If we can't do it it can't be done."

CASE 7. I had been working with a development man in one of the three largest hospital supply companies in the country. He had been much interested in a very simple product that we had that would be part of a system in which the company had a major, but not dominant, share of the market. He was quite interested in the product, had conducted most of the tests necessary to show the superiority of our product over what was being used in his company, and was about to recommend its acceptance. He was unexpectedly transferred, a promotion, incidentally, to another department. He referred me to another individual to carry out the project and to discuss other products of ours.

I made an appointment with the new individual who had the title Senior Development Engineer. Anticipating no difficulty because of the interest and past work, I arrived, prepared to demonstrate our accomplishments. I was kept waiting in the lobby of the building for twenty minutes. I was then met by the Senior Development Engineer. We went, not to his office, but to a crowded cafeteria, where he asked me to explain what I wanted. At a table for eight, with six other non-involved people who were not only uninterested but rather irritated because I was interrupting their coffee break, I demonstrated our product that, incidentally, had been made part of his company's commercial system by the development group that I had worked with before. The first question I was asked was how much it would cost. I was told, not bluntly but honestly, that the company was not interested in anything new. That is an exact quote. I was further told that, regardless of the advantages of our product, it would be judged only on whether or not it had a price advantage over the present product. The rest of the quote about not being interested in anything new was that they would prefer to see someone else use the product for a period of time before introducing it into their line. This, in spite of the fact that it was a patented product that would be given to someone exclusively, and in spite of the fact that another part of his own company had tested it and found it satisfactory. And this from a company that, in 1980, reported an expenditure of almost $60 million of its money for research and development.

What have I learned from these experiences? Not much about research and development and technology. But a great deal about human relations. I learned that very few companies have much appreciation for what could only be described as common courtesy. I don't think they are consciously discourteous. But the type of arrogance exhibited certainly gives the impression of a lack of any consideration for those outside the organization. The seeker of funds cannot afford to be aggressive. The unanswered phone calls, the broken promise to "call you in a couple of days," the samples accepted for test that remain on the desk for months or that disappear without a trace into the great maw of the system are commonplace.

In considering the almost universal uniformity of such action I can't help reflecting with a somewhat guilty feeling on the many years I spent on the other side of the table from where I am now when I, too, was considering suggestions from outside my own large organizations. And I wonder whether or not I, too, was guilty of such insensitivity. I don't think I was. I don't think so because, all my life, I have admired creativity and creative individuals. All my life has been spent with creative individuals. But if in some moments of conditioned involvement in the establishment I have turned off the creative person I hereby apologize.

If inventors somehow have succeeded in starting a company, have production and sales facilities, and then require more financing they are still in trouble. More small businesses fail because of underestimation of the financing required than for any other reason. The objective of any entrepreneur is to *get* the company going. Too often he or she loses sight of the necessity to *keep* it going. The average entrepreneur has no concept of the peripheral costs of establishing a business and maintaining it past the start-up stage. Each knows it will cost something for production and production equipment, and can estimate that reasonably accurately. The entrepreneur thinks he or she can estimate what further development work will cost, but usually these estimates are based on what the entrepreneur personally can do, not on what the results will be if there is dependence on others. The cost of having technological problems solved outside your own organization is usually much greater than estimated. The time is always longer. My own rule is to estimate what I think should be a reasonable time to get results, and then double it.

Because of the highly technical nature of some of the work done by my company, I had started the practice of having work done by the most highly qualified group available, wherever in the world it existed. When a prototype instrument that I had commissioned abroad arrived at the airport, I received a call saying it was in customs. Rather naively, I said I would pick it up. It was then that I was told that, although the product was of no commercial value, it had been evaluated at more than $250 and, therefore, I would have to have a customs broker get it through customs. Following is the cost of paying the broker for clearance, and for delivery of the small package to my office:

Freight Charges		$ 95.73
Duty Deposit		110.24
Bonds to Customs		20.00
Dock Storage		5.00
General Order Warehouse		45.00
Release from General Order Warehouse		6.37
Cartage		27.50
Reforwarding Services		2.50
Services for Customs Examination		4.50
Messenger Service		7.50
Telephone		1.00
Brokerage Fee		65.00
	Total	$390.34

Everyone knows that legal costs are high and probably necessary. It is rare that actual cost predictions for a start-up business are made accurately. In my first years of corporate existence legal fees were the highest single expenditure of my company—even higher than research and development. In order to manufacture and market a few of my products I established a Limited Partnership with my company as the General Partner and several individuals as limited partners. The resulting document required 78 pages of legal-size paper for its description. In addition, there were the state and federal forms to be submitted, and following approval, the endless data to be collected for the quarterly reports, and the periodic questionnaires to be completed.

Two groups of lawyers were involved in the formation of the Partnership—one to draw up the legal documents, and the other to represent the partners. At the end of the first years of operation the auditing firm (another necessity) wrote to the lawyers asking two questions: (1) Did the Partnership owe them any money, and (2) Did they know of any litigation or pending litigation against the Partnership. Following is the answer in full provided by one of the firms:

> At the request, dated January 28, 1980 (the "Letter") of Dr. Thomas P. Carney, President of Metatech Corporation, which is the general partner of Metatech Hospital Products, Ltd. (the "Partnership"), we advise you as follows in connection with your examination of the accounts of the Partnership.
>
> This firm did not represent the Partnership in matters pertaining to its formation and the sale of limited partnership interests. Subsequent to its formation, we have represented the Partnership in specific matters which have been referred to us, primarily with regard to the admission of additional limited partners to the Partnership. There may, therefore, be matters of a legal nature about which we have not been consulted.
>
> This response is limited by, and made in accordance with, the American Bar Association Statement of Policy Regarding Lawyers' Responses to Auditors' Requests for Information (December, 1975), including the Commentary thereto, which is an integral part of that Statement of Policy. Without limiting the generality of the foregoing, the limitations set forth in that Statement on the scope and use of this response (Paragraphs 2 and 7) are specifically incorporated herein by reference, and by description herein of any "loss contingencies" (as that term is defined in Paragraph 5 of that Statement) is qualified in its entirety by Paragraph 5 of that Statement and the accompanying Commentary. The terms used herein have the definitions given them in that Statement.
>
> Subject to the foregoing, and to the balance of this letter, we advise you that since January 1, 1979, we have not been engaged

> by the Partnership to give substantive attention to, or represent the Partnership in connection with, material loss contingencies coming within the scope of clause (a) of Paragraph 5 of such Statement. To determine materiality, we follow the definition provided in Item 5 of Regulation S-K under the Securities Act of 1933, which uses a standard of 10 percent of current assets on a consolidated basis, permits the omission of certain ordinary-course litigation, and requires inclusion of certain environmental proceedings, regardless of monetary standards.
>
> The Letter does not specifically identify any loss contingencies coming within the scope of clauses (b) and (c) of Paragraph 5 of such Statement. Accordingly, consistent with such Statement, our response is limited to loss contingencies within the scope of clause (a) of such Paragraph 5.
>
> On December 31, 1979, there was approximately $300 accrued, but unbilled, for services rendered during the period ended December 31, 1979.
>
> This information is furnished as of the date hereof. Except as otherwise noted, we assume no obligation to advise you of changes which may hereafter be brought to our attention.

I have a sneaking feeling that the $300 accrued by the firm but unbilled was for part of the cost of developing the obfuscatory language in the letter. Obviously the reply was not meant to inform. It was meant to protect the lawyers involved.

A company that is operating is an ideal candidate for consideration by the venture capitalist. The decisions that must be made do not concern the product, but only whether or not the product can be sold. There are already data to be analyzed based on past sales experience. So the analysis lies within the capabilities of the venture capitalist group.

But it is here, too, that entrepreneurs are most likely to lose control of their companies. It is probable that they have invested all that they own in the operation. Their entire lives are involved in making the company a success. Therefore, they are likely to make whatever compromises that are necessary to continue operation. Thus, if it is necessary to give up control of both the financial and technical aspects of their companies in order to have them survive, they are willing to do so. Possibly it is wrong to say they are willing. They will do so because survival depends on it.

And what happens then? Then the company has a good chance of surviving. It will survive because whoever has supplied the money has decided that there is no risk in operating the company under conditions laid down by the financiers and not the entrepreneur. It will also survive, not as a risk-taking venture that has the opportunity for technical advances, but as a company that will produce and sell a product that might not cause a great leap forward for humanity, but that will make a profit.

In dealing with a professional investing group on programs relating to high technology products, the technical evaluation and advice available to the investors is of primary importance. They do not, and most emphatically should not, depend only on the data presented by the inventor or the inventor's advisors. In many cases groups established for investment, or the so-called "risk capital" groups, have their own in-house expert. They usually assume that one such technically trained person is enough to advise numbers of economists and investment analysts. Because of their training these experts are held in high esteem. Their power is great. But, truly, each is the one-eyed man in the land of the blind. The first sign of technical doubt from the expert immediately causes all sorts of warning flags to run up the flagpoles. It is then that the symptoms every inventor has come to recognize and to dread appear. The evaluators begin to search for reasons why a project should be rejected, rather than why it should be accepted.

The in-house people are the experts. Therefore they assume to themselves a great deal of knowledge. Whether it is chemistry, physics, electronics, manufacturing—anything that relates to that generality called technology they feel they must give a decision on since that is what is expected.

I have been impressed with very few of these in-house experts. Possibly it is uncharitable to say so, but I have the feeling that they are in their jobs because the organization recognized it needed technical advice, but confused exposure to a technical education with technical knowledge and ability. It is even more uncharitable to say that, if they really had the knowledge and ability to do what their financial friends thought they were doing, they would have jobs where their ability would result in their own creative developments.

The in-house experts are expected to evaluate, and they do. However, there is a great similarity between their action and those of a government regulatory agency such as the FDA. If they react positively, there is the chance that they would make a mistake, and an investor would lose money. This would reflect very badly on the group. If, however, they react negatively, they still might be making a mistake but no one would know about it. On the contrary, the group would be credited with being cautious with investors' money.

And that is another reason why there is very little risk capital available. The failure to provide adequate capital to allow small firms to operate has had two effects in recent years, both bad in my opinion. Senator Gaylord Nelson, Chairman of the Senate Select Committee on Small Business, has called attention to the rapidly increasing investment of foreign firms in U.S. small businesses. He mentions specifically that a West German firm paid 73% above the market price to gain controlling interest in Advanced Micro Devices, a Sunnyvale, California company manufacturing sophisticated electronic equipment, and that Fujitsu, a Japanese company, purchased 28% of the stock of Amdahl Corporation, also of Sunnyvale, a small computer manufacturer, because Amdahl was unable to raise American capital. He further points out that the Senate Small Business Committee has recorded fifteen such transactions having taken place between August of 1977 and March of 1978.

The Conference Board reported that, in the second quarter of 1979 there were 35 announcements of foreign acquisitions of U.S. manufacturing facilities. In 1978, for the same period, there were 41 such acquisitions. In the fourth quarter of 1979, 60 acquisitions were made by foreign companies, while in the first quarter of 1980 there were 39 acquisitions.

Probably the most significant result of such foreign takeovers, in addition to loss of U.S. profits, is the access it gives a foreign company to U.S. technology. Ownership in a corporation obviously conveys the right to ownership of whatever new developments take place. Thus, there is a direct transfer of technology from creative U.S. industry to countries abroad. With such a situation existing it is no wonder that our technological lead over the rest of the world is declining.

In addition to the foreign takeovers of small business, more than 2,000 small enterprises in the United States were acquired by large companies in 1976 alone. This, obviously, is a recognition by larger companies of the value of small business. But it doesn't make much sense to acquire a business when the very act of acquisition destroys the quality that made the acquired company desirable in the first place, namely, independence. The value of small business to a nation has been recognized, although not generally appreciated. The Small Business Committee reports that in 1978 small business accounted for 43% of our Gross National Product, and employed more than half of all workers in the private sector.

In particular those small businesses based on technology are important. Technology in general has contributed more than its share to the economy of the country. According to a 1977 Commerce Department study, technological innovation produced 45% of the nation's real economic growth from 1929 to 1969. High technology firms created jobs 88% faster between 1957 and 1973, and their productivity grew 38% faster. High technology exports have produced the only positive figures in the United States balance of payments in recent years, except for agriculture.[1]

A report prepared in 1978 by Jacob Rabinow of the National Bureau of Standards at the request of the Office of Management and Budget gives official credence to the increased efficiency of research groups representing small organizations. The report has not been published by the department. In it Rabinow points out that small businesses accounted for almost one-half of the major inventions between 1953 and 1973, that the ratio of innovation to sales is about one-third greater in small firms, and that small firms produce four times as many inventions per R&D employee as do large firms in spite of the fact that it costs a small company only half as much to support a researcher as it does a large company.

He also points out that only 3.5% of government research contracts to all R&D institutions, that is, industry, in-house laboratories, education institutions, and federally financed R&D centers, went to small technology based groups.[2]

In 1967, a panel on Invention and Innovation chaired by Robert A. Charpie reported on the status of innovation in the United States. The Charpie Report refers to the following studies that emphasize the importance of independent investors and small business: Professor John Jeukes showed that of 61 inventions and innovations of the twentieth century over half came from independent inventors or small firms. Professor Daniel Hamberg of the University of Maryland studied major inventions made between 1946 and 1955, and found that over two-thirds came from outside large organizations. Professor Merton Peck of Harvard studied 149 inventions relating to the aluminum industry.

Only one of seven inventions came from a major company. He also selected 13 major innovations in the American steel industry. Four came from European companies, seven from independent inventors, and none from American steel companies. Professor John Enos of MIT studied seven major inventions in refining and cracking petroleum. All seven came from independent inventors.

Simply listing some of the major innovations of the 20th century that originated in small firms or came from independent inventors will give some idea of these contributions to our economy. These include the vacuum tube, the cottonpicker, the zipper, xerography, automatic transmissions, FM radio, the Land camera, the mercury dry cell, air conditioning, Bakelite, and the oxygen steel-making process.

An interesting example of the lack of appreciation for both innovators and entrepreneurs in companies, and the consequences of this failure, is given by Ed Roberts of MIT. He studied a large electronics system corporation in the Boston area. He traced forty-four of its former employees who left to participate in forming thirty-nine new companies. In 1966, thirty-two of the thirty-nine companies were still in business and their combined total sales amounted to $72 million. The striking thing is that, in the same year, the organization from which they came did only $30 million worth of business—less than half the sales volume of the spinoff companies.

The independent inventors, the entrepreneurs, those operating small businesses will survive. They will survive in spite of all the obstacles and road blocks strewn in their paths. They will survive because of an indomitable will that forces them to continue. But, considering their importance to the future, there should be some way to make it easier for them to survive, to encourage their survival, and to increase their effectiveness.

Because great contributions have come from independent inventors and small businesses due to their willingness to make personal sacrifices, the idea seems to have developed that one of the requirements for success is sacrifice. The individuals have become heroes, and there is something noble about the fact that they have endured adversity.

I would say that sacrifice is not a necessary ingredient for success, but, given our present environment, it seems to be unavoidable. Given proper support, the creative individuals can be even more productive when they are not enmeshed in the details of raising money, in convincing reluctant backers of the value of their work, in overcoming skepticism as to their ability. I know—I have been comfortable and uncomfortable. Comfortable is better.

8 . . . wherein the world conspires against the independent inventor, but the inventor is important and the results of creativity will be recognized

The life of independent inventors is not a comfortable one. They will spend their time submitting their inventions to anyone who will listen, and then be rebuffed. They will spend their savings, and mortgage their possessions to continue their work in the firm belief that eventually someone will share their enthusiasm and recognize their genius. The plight of independent inventors is not a happy one. The whole world conspires against them. The tax system, the patent system, the industry—all are united in an effort either to prevent them from inventing or, if they overcome all the obstacles placed in their path and do succeed in inventing something, to prevent them from marketing it or, worse, to steal it. Or so it seems to the inventor.

The views of independent inventors are not modest views. They never invent anything of *little* value. In Boswell's *Life of Johnson* it is recorded that when Johnson, as the executor of the estate, was selling Thralis's brewery and was asked the value of the property he responded: "We are not here to sell a parcel of boilers and vats, but the potentiality of growing rich beyond the dreams of avarice."[1] All inventions, in the eyes of their inventors, have the potential to bring riches beyond the dreams of avarice, and the inventors are always puzzled when those less gifted than themselves do not recognize the promise.

And yet, if the combined real genius of all the independent inventors could be harnessed, if there were some way to select those ideas and then to develop them, we probably would not be talking about the loss of a technological advantage for this country.

In many ways the world does conspire against the inventor. The tax system is one of the discriminatory means used to make life difficult. A company, large or small, can take the expenditures for its research and development as a tax deduction. Thus, the government pays for about 50% of the research done in industry. Further, if the company has one successful product and several unsuccessful research programs, the losses from its failures can be deducted, and so offset some of the profits of its successes. Unless independent inventors become companies they have no such favorable tax treatment. Everything they spend—patent application, materials, equipment, prototype production—all are total losses, not deductible from their taxes. If they have two programs, one successful and one unsuccessful, they pay taxes on the profit of the successful product, but they cannot deduct from these the losses sustained by other programs.

Probably the only device that inventors have to protect themselves against the theft of their ideas or products is the patent system. Yet the system has become so cumbersome and expensive that it is almost beyond the reach of an individual, particularly if foreign patent coverage is needed. It requires about two years and about $5,000 to have a patent prosecuted through the United States Patent Office. The time is no different—in fact it is usually less—for foreign patents, but the expense, involving mostly attorney's fees and translations, is a multiple determined by the number of countries involved. In spite of this, about 16,000 patents are issued each year to independent inventors.

Another troubling legal aspect involved is the attitude of the courts toward patents. Almost routinely, 80% of the patents that are in litigation are declared invalid by Federal judges. A patent is granted after thorough review by an examiner who usually has both scientific and legal training. Yet the judge, who probably without exception has very little technical knowledge, sees fit to overturn the patent department's judgment. This, of course, is a problem for corporations also, but it is more critical for independents, since usually they must depend on patent protection as a condition for licensing, and they do not have the financial resources to do battle in court.

Industry has two ways to discourage independent inventors. The first is caused usually by the legal department which resists any communication between an inventor and the company. There is some justification for this, although not to the extent of its practice that approaches paranoia. The fear of the company is that an inventor will submit an idea or a product, have it refused, and then, some time in the future, see the company produce the same or a similar product that it claims as its own. The natural tendency is for the inventor to sue, claiming the company stole his or her idea. In very rare instances such dishonesty might have taken place. However, it is more usual for a company with a large research staff to have had the same idea at the same time as the inventor, or even before the inventor. Products are usually based on needs and, if a need is apparent, it is reasonable to suppose that the filling of that need will become apparent to more than one person. Whatever the situation, the legal mind decrees that such suits will be prevented by eliminating any possibility of such a situation arising even though by so doing it might also eliminate the possibility of being able to consider a revolutionary idea. Thus, many large companies have the policy of returning to the inventor, unopened, a communication regarding an invention if it can be so identified. If the letter is opened, it is returned with a certified statement that it has not been read.

Another legal requirement of most companies is that the inventor sign a so called "confidential disclosure" agreement. These agreements usually have the effect of convincing the inventor that the company really does want to use him or her with no compensation benefits. Also, usually, as they are designed to do, the agreements discourage inventors from bothering the company with their ideas.

Three such agreements, selected at random from my files, will illustrate what I mean. All are from large, reputable companies.

The first has these provisions:

> No obligation of any kind is assumed by, or may be implied against company or until a formal written contract has been entered into, and then the obligation shall be only such as is expressed in the formal written contract.
>
> I do not give company any rights under patents I now have or may later obtain covering my suggestions, but I also hereby, in consideration of its examining my suggestion, release it from any liability because of use of any portion thereof, except such liability as may arise under valid patents now or hereafter used.

A second agreement has these provisions:

1. Company is not obligated to study the idea.
2. Company shall have the right to retain any material submitted to it, to make copies thereof, and to retain the same in its files.

3. The agreement does not give company any right of any kind in the ideas, new devices, or improvements submitted to it.
4. No obligation of any kind is assumed or implied by company unless or until a formal written contract has been entered into, and then the obligation shall be only such as is expressed in the formal written contract.
5. In the absence of a formal written agreement between company and the submitter, all rights and remedies of the submitter of ideas, new devices, or improvements (and principles, if any, thereof) arising out of the disclosure to, or the use thereof by company or any of its representatives shall be limited to any rights and remedies as may now or in the future be accorded to the submitter under U.S. patents or copyrights.
6. All claims of any nature whatsoever arising out of any disclosures by the submitter to company are hereby waived, except such rights as may hereafter be set forth in the formal written contract referred to in Paragraph 4 hereof, or rights and remedies now or in the future accorded to the submitter under U.S. patents or copyrights.

The third is a short, informal letter that says:

> You agree that the submission of your suggestions or ideas (as well as any additional ones which you may hereafter submit as incidental to the material already submitted) and any consideration which may be given to them at your request by company shall be without obligation of any kind assumed by or implied against company unless or until such is expressed in a formal written contract between us.

Is it any wonder that inventors might think they are not loved by the large company? When the paranoia (in inventors it seems to be an inherited gene trait) of the inventor is reinforced by the paranoia (in companies it seems to be an acquired characteristic that comes with bigness) of the company there is not much chance for an exchange of ideas.

When a company establishes what amounts to a legal screen between itself and new ideas it has made the decision that it has within itself all the creativity necessary for growth. It is a rather arrogant attitude, since no organization today has a monopoly on brains.

For about thirty years I was in the position of being able to consider the ideas of outside inventors. For the past five years I have been on the opposite side of the table, offering ideas and products to others. In neither of these situations did I pay much attention to the formalities of legal letters of confidentiality. When I was representing large companies and an inventor signed a letter of agreement, I accepted it because it made the legal division happy. If he or she did not sign, I did not let that interfere with allowing a presentation of an idea. In all that time my companies were never sued by an inventor. And I did get a lot of good ideas.

Industrial organizations have an additional protection from accusations of pirating ideas in their normal record keeping procedures. In most companies research people are encouraged to record a "concept of invention." This means that when a person has an idea for a new product he or she records it in as much detail as possible. It is signed and dated, and the entire document is read and witnessed by another person. It is then filed, usually in the legal department. In addition, each research worker is required to keep a notebook detailing on a daily basis every experiment that is performed. This, too, is signed, dated, and witnessed. Thus, if a company really has been working on an idea, or has even thought of an idea that is presented from outside, it should have ample evidence to satisfy the independent inventor as to the priority.

I have expressed elsewhere my feeling that a new idea is a very precious thing. I believe it is worthwhile taking the chance on being sued if the result might be new ideas. Suits can be eliminated by preventing new ideas. If the company is in a firm position, it is creating the atmosphere in which new ideas are encouraged, with the price being a possible lawsuit. I would trade a few lawsuits for a new idea. In addition to having the idea, the suits might even keep the legal department busy at some reasonably nondestructive work.

When, in my present company, I in effect put myself in the position of the independent inventor, my opinion regarding the inventor–company relationship did not change. In all my years in industry I was not aware of a single incidence of a large company stealing the idea of an individual. Consequently, I have been quite open in presenting products to other companies.

I have not hesitated to sign even the most outrageous confidential disclosure statement. I believe that practically all established companies are honest, and that I will receive the same treatment with or without a signed letter. Since it doesn't make any difference to me I might as well make a bean-counter happy by conforming to the system.

In addition to my belief in the fundamental honesty of technical organizations, I have another reason for believing that it will be of no disadvantage to me to bend over backwards to maintain contact with large companies. It is a very pragmatic reason, and I am probably as susceptible to the accusation of arrogance as are the organizations who say they don't need outside ideas. I don't believe a large company can afford to steal one of my ideas. A company, too, has a reputation to uphold. It cannot afford to be accused of dishonesty by a reputable inventor or organization and, with the experience I have had, I think I am in a position to judge whether or not I am being dealt with ethically.

Unfortunately, the average independent inventor does not have such an understanding of the ponderous workings of a large company. He or she sees only that the organization is making it as difficult as possible for the inventor to communicate with it.

If inventors do happen to break through the legal curtain they are apt to encounter an even more frustrating experience. It is then that they come into contact with the ubiquitous attitude known as the NIH—Not Invented Here—syndrome. It is doubly frustrating because now obstruction is not caused by lawyers, whom the inventors are not expected to understand, but from their colleagues, their fellow workers in the scientific field. It falls logically to the responsibility of the technical division of a company to evaluate the efficacy of an invention, whether it comes from inside or outside. If an idea is not technically good, almost without exception the right decision is made. Very few bad ideas are accepted for development. The difficulty arises when an idea comes to an industrial organization from outside in an area in which the research group is already working, and the idea appears good. The objective all too often seems to be to find some reason, logical to the technical people and to management, for rejecting it.

Please believe I am not imputing any evil motives to the research groups. It is an unconscious reaction. The group is probably composed of first class scientists, being paid good money because they are expected to invent. It is almost a loss of face, therefore, to have to admit to their management that some individual outside appears to have done what their high pressure group did not do. Notice that I said "did not do," not "could not do." I have made the point elsewhere that organized research programs presently do not have flexibility to think. They are more often than not defending and improving what they have, not expanding their horizons.

The National Industrial Conference Board, in its recent study, "Perspectives for the 70s and 80s," reported on a study among middle management in twelve major U.S. companies. The largest proportion rating their own company "outstanding" on its receptivity to change was only 49%. In another study, people at different job levels were analyzed in terms of their attitude toward change. The startling finding was that even among scientists and engineers, only 20% were enthusiastic and deliberate promoters of change, while 70% were apathetic and 10% resisted change of any sort.

Another difficulty is the fact that the failure of an inventor working on an outside contract is much more visible at the corporate level then is the failure of a project being worked on as part of an overall internal program. The failures in internal programs are assumed to be normal. However, it seems to be difficult for people outside of the technical area to accept failure of an outside project. In the first place, the amount of money being paid the inventor, either for work on the process or rights to the product or process, is known to everyone. There seems to be something much more serious about a known million dollar outside project failing than there is in a million dollars worth of overall general research failing.

When I established my own company it was with the objective of developing products based on high technology. I decided we would not develop a product unless it served a need, and unless it was either unique or was demonstrably and appreciably superior to anything else in the same field.

For many years I had been in the position in large organizations of considering inventions brought to me from outside. I thought I knew what was necessary to produce the evidence on which a favorable decision could be made. Many times I solemnly wished that an inventor knew both more about his or her own technology and also about the way a business operated. So when I knew I would be in the position of the inventor and not the company decision maker I decided that I would not present to any organization a product that had not been developed at least through the prototype, that is, the first working model, stage. I did not have to do too much profound reasoning to conclude that it was easier to sell a product than to sell a concept. Concepts fail more often than products. Therefore, to establish and maintain my credibility I decided to take the risk of making my own mistakes in private and at my own expense. As a plan, we decided to work for up to four years, if necessary, developing a whole line of products before we even told anyone they were available.

In spite of the fact that I have adhered to this principle, I, too, have had the experience of having good projects rejected.

So independent inventors must pass three categories of obstacles. First, they must invent, overcoming tax discrimination and obtaining sufficient finances to pay for their materials, equipment, and legal fees. Then they must get someone to listen to them. Finally they must obtain approval for possible further development and marketing and they must obtain financial investment in their products from a group who considered them an intrusion in the first place.

There is available another source of ideas, and that is the universities. In the science departments are some of the most creative people in the world. They, too, could be classed as independent inventors, although their inventing is done almost incidental to their academic work.

The problem is to arrange the proper cooperation between the industrial and the acacemic groups in order to make practical the ideas of the academician. The academic scientist has what industry needs—ideas, and the ability to reduce an idea to an initial practice. Industry has what the academician needs—the resources to take the initial product through development, production, and sales.

Why haven't they cooperated more often?

There are many examples of successful cooperation. Insulin and liver extract, Merthiolate, the Cephalosporan antibiotics, Benadryl, streptomycin, all came initially from academic laboratories. Cooperation between the organic chemists in universities and the pharmaceutical industry is a natural one. Industry has a vast potential for screening chemicals in animals. Academicians usually have no such facility available. When, in the course of their work, chemists synthesized a large number of compounds, it was natural that they should submit them to whatever screening procedures were available. Most pharmaceutical houses made their screens available to these scientists. If a promising lead developed, the company involved would negotiate the right to develop it into a marketed product.

One of the changes that has come about that has made this

arrangement less attractive to industry has been the increased government support of academic research. There are very few research people in academic laboratories that do not have some government support. With government support also comes government control. So, if a patentable product arises from research sponsored by the government, the patent must be assigned to the government or, at least, the government must have a right to license the patent. This immediately destroys the incentive of a corporation to invest in the development and establishment in the marketplace of a product. No company would spend the time and money on such a product unless it could be assured of some exclusivity to enable it to get back its investment before competition can take advantage of the information that has become public knowledge.

It is unfortunate, this terrible waste of creativity. Those working outside organized research groups are perhaps the best source of totally new ideas. Organized research almost by definition now means organized thinking. And organized thinking, almost by definition, eliminates thinking anything different. One of the things organization does is to *organize out* individual creativity.

Outside inventors have at least two advantages over company inventors. In the first place they are not bound by time as are the company employees. They are bound only by the degree of their persistence, and the degree to which they are willing to sacrifice personal comforts to the belief in their success. A company cannot, of necessity, make such sacrifices. It must make a decision on whether or not a project will bring in a revenue in some reasonable period of time.

The second advantage is that inventors are, by choice, totally free souls. They are not bound by company policy, nor are they aware of the technical difficulties involved in producing products. Therefore, they are free to work on ideas that an organized research group would have to reject.

If it is true—and I am totally convinced that it is—that independent investigators can contribute in a major way to industry, and if it is true that the inventors are not welcomed into organized research groups, and if it is true that the only reason for the lack of a warm reception of inventors is that they increase the discomfort index, then attempts should be made to find ways of making them welcome without antagonizing the organization.

When any company first begins research, everything that it does can be original and creative, since it need not make its projects fit into an already existing program, or have its problems determined by the special abilities of already organized research people. However, as time goes on, and as products result from research, it becomes necessary to expend technical effort on defending these products. I recognize this as a necessary activity. However, the degree to which an organization concentrates on what I call defensive research to the exclusion of research on totally new products will determine the creativity of the company.

An improvement in an existing product might be of benefit to a company since it will allow the company to retain or increase its market share. The expenditure of the same research effort might result in a totally new product in an area that would expand the total potential. However, it is usually safer to try to improve a product than it is to find a totally new product. Thus, an increasing effort will go toward changing what is already available rather than doing something totally new.

It is doubly unfortunate that the independent inventors are not given more recognition, since it is possible they could solve one of the problems facing big technology industry. If industry continues to insist on immediate profit with a minimum of risk, it, as I have said, eliminates long term projects where the results of research might be unpredictable. However, where an inventor can place before a company a product that can be examined and evaluated and when, in many cases, the development has reached the stage where part of the production problems have been eliminated the biggest barrier has been overcome. In most cases research and development are not understood by general management. A factor that might be totally insignificant to researchers because they understand the problem becomes a major stumbling block to management's acceptance of a product because they don't understand it. And, if the objective is to eliminate all risk, what is not understood is perceived as a risk. With the opportunity to examine a finished product the risk of research and development is no longer present. The problem facing management, then, is whether or not the product can be sold at a profit in the marketplace, a problem with which they are familiar and comfortable.

Independent inventors are not the only inventors seeking recognition. Even within industrial organizations the originators of new products sometimes feel that they are not being rewarded adequately.

It has been suggested that both as a matter of equity and as a means for further motivating a worker, a scientist be allowed to participate in the profits resulting from anything he or she might invent. When scientists begin work in an industrial laboratory they give up all rights to anything that might result. If a patent issues on the work, it is assigned to the company in exchange for a small payment, sometimes as little as one dollar, and "other considerations." Even if the company makes a million dollar profit as a result of marketing the invention the inventor still does not participate directly in any of that income. On the surface it doesn't seem quite fair. There has been some agitation from scientific organizations for change, and there have even been suggestions that the situation be changed by law.

However, having had experience both as a laboratory worker and as an administrator I am strongly opposed to the idea of inventors participating in profits from sales of individual products. Consider first what some of the "other considerations" are that inventors receive in exchange for the assignment of a patent to a company. First of all, they are furnished with a place to work. They have available the use of millions of dollars worth of scientific instruments, and staffs of specialists who perform service work for them. They usually have assistants to help them with their own work. They have library and information collecting help. And, in addition, they have a salary, fringe benefits, and a retirement income. All of these things these scientists have even though they may never invent a marketable product during their entire careers.

If something positive does result from their work the company is then faced with the task of making it practical. This involves investment in developing a manufacturing process and in an advertising and sales program. To put it on a risk-comparison basis, the company is risking some millions of dollars on any product, while the scientist is risking nothing and is, in fact, guaranteed a salary and other benefits.

I don't want to indicate that the productive inventor is not rewarded above the unproductive one. The rewards come as a result of increased salary, stock options, and other benefits not directly tied to individual sales.

There is another very important administrative reason for

not paying company inventors royalties. Usually scientists are not totally free to work on problems of their own choosing. They work on problems that are of benefit to the company. In many cases they are assigned specific projects, and in some cases these are projects from which patentable inventions can never result. Nevertheless, they are of importance to the entire research program, and contribute to its overall success. Obviously it would be unfair to assign one worker a project from which he or she could not benefit financially while, at the same time, assigning another a project that might give a royalty return. For all practical purposes it would probably be impossible to get anyone to work on basic research problems or on development problems under these circumstances, even though they are essential to progress.

Another difficulty arises when an attempt is made to assign credit for contributions to the overall success of a product. Granted, an idea, an invention, must come from one mind. But someone must turn that idea into a practicality. It is not unusual for a technical contribution, made in production for example, to transform an economically impractical idea into a price-competitive product. In view of all the technical input necessary to carry an idea to the market it is almost impossible to assign degrees of importance to each individual so as to reward him or her for contributions to a specific product.

For all these reasons, then, it seems to me that for the sake of fairness, equity, and motivation of creative people, it is inadvisable to reward individual inventors by giving them royalties on their products.

Of course, if an inventor wants to assume the risk of doing without salary, supplying his or her own equipment and instruments, and making a first prototype of a product, the situation is then totally different. At that point the company must make the inventor a partner in its sales. In this case, the inventor has assumed the risk formerly taken by the company. And in this case the company must pay for the elimination of that risk. But inventors cannot have it both ways. They cannot demand the security given by industrial employment and, at the same time, expect the rewards of the independent inventor. If they want freedom they must pay for it.

If company management has difficulty understanding the inventor's research and development process, the inventor finds it equally difficult to understand some marketing decisions. It is in the nature of all inventors, whether they invent a potato peeler or a machine to cure cancer, to believe that the product of their minds is the greatest advance of the century. It comes as a shock, then, to be told that there is no interest in marketing their product. What inventors fail to realize is that a technological triumph, regardless of how brilliant it may be, is not always saleable. A product must fill a need, or must be one for which a need can be developed. And the judgment as to whether or not there is a need, that is, whether or not the company can sell the product must, probably fortunately for the well-being of the company and the inventor, rest with the company.

Inventors never accept a negative decision on their products. In addition to their unshakeable confidence in themselves, they are aware of enough examples of developments—such as Xerography, the vacuum tube, and the mercury dry cell—that have turned into revolutionary products to encourage them to persist in their efforts to have their inventions accepted.

The introduction of the Bessemer Process for making steel is a classical example of how a new idea is resisted, despite its advantages over traditional practice. Henry Bessemer's new method of steelmaking was looked at but ignored by the ironmakers of the mid-19th century. Why? Says historian Elting Morison: "Few men look forward cheerfully to even a perfect revolution—the converter introduced a new element with strange characteristics and undescribed energies."[2] What finally changed this? The consumer. Says Morison: "The railroads were the nurse crop for the steel industry."[3] The impetus did not come until the railroads demanded steel. Meanwhile, years had elapsed since Bessemer's idea had been introduced.

It should not be assumed that it is only with industry that the inventor has difficulties. Professional groups established for the express purpose of investing in new ideas suffer from the same lack of imagination, ability to evaluate, and willingness to take risks as do their industrial counterparts.

I have sounded pessimistic in some of my analyses. But, strangely enough, I am not pessimistic about the ultimate result. Entrepreneur inventors must be willing to expose themselves to many disappointments if they are to see their products through to successful marketing. Some of their difficulties are caused by their own inexperience and overenthusiasm. Most individual inventors are undercapitalized, and the underestimation of capital required is usually a direct result of overenthusiasm as to how fast positive results will be obtained. It is inconceivable to inventors that it should take months for someone to decide that their product is good. The committee system, the legal departments, the marketing research department, and finally the sales department—inventors consider them simply irrelevant if the product is good technically. But they come to tolerate, if not to understand.

Inventors are like horse players or gamblers. If they lose today, there is always tomorrow. So they sell their stock, they spend their savings, they borrow on their insurance, they mortgage their houses, all to keep going until the time that someone recognizes their genius.

They must also be willing to suffer some indignities along the way. They don't have to like them but, for the sake of their ventures, they must accept them. Arrogance cannot be part of their reaction. That is a luxury reserved for the dispensers of the money. Their technical ability will be doubted, particularly by those seeking excuses to reject rather than accept the proposals. And, in some by no means rare instances, even their integrity will be questioned.

But always there is the possibility that they will have the next Xerox.

It is the fate of most inventors that they remain anonymous. There are obvious exceptions. Probably everyone associates Land with Polaroid, but this is because of his continuous association with publicity for his company developments. The name of Jonas Salk is a household word, but this is because of the program of publicity associated with the March of Dimes.

But who ever heard of Frank Colton? Dr. Colton's name is on the patent, he is the inventor, of the first birth control pill. Do the names Horace Shonle and Wilbur Doran mean anything to you? They revolutionized a part of the practice of medicine when they invented the barbiturates Amytal and Seconal. Harvey Abramson is the inventor of record of the copper interuterine device but, except for the publication of the patent, I don't believe his name has ever appeared in print in association with this invention.

One of the reasons for the lack of recognition of the original inventor of a product is the time involved between conception of an idea or the invention and the actual proof of effectiveness or practicality that converts the idea into a marketable product.

By the time a product is marketed the inventor in many cases has lost contact; certainly he or she has lost control over activities related to commercialization. This is particularly true in the invention and development of drugs. The patent on the product in the first birth control pill was issued in 1955. This means that the invention probably took place in about 1952. Yet the product was not approved for marketing until 1960. In that length of time the original inventor might have worked on three or four other, even unrelated, programs.

When a major new drug is announced it usually comes from a clinic, a hospital, or a medical school, and names the clinician who is reporting the results of his or her trials on human patients. I suppose this is natural, since the objective is to find a product useful in people, and the clinician is the last link in the testing chain. Clinicians are an absolutely vital part of the development procedure. Yet, by the very nature of their work they cannot do anything creative. They can observe new and important actions, either good or bad. But by the time they are given the drug for testing, all the creative, even the innovative work has been done. The product has been either synthesized by a chemist or isolated from nature. It has been tested in animals, and toxicity and physiological action have been determined. Pharmaceutical forms have been prepared, and the stability of these forms studied. Production methods have been developed, and quality control procedures have been established. So the clinician is asked, in effect, to confirm or deny observations already made.

It is a traumatic experience—to have done something creative in research, to know it will result in a significant advance in medicine or in any other field, to have to relinquish control over all those things that can make the invention succeed or fail, and then, when it becomes a reality, to have been disassociated with it for so long that it seems necessary to reclaim the whole project as one would reclaim a child of one's own that had mistakenly been put up for adoption.

Research for those interested in having the results of their work used to increase the quality of life is traumatic. But I don't know of many people engaged in such work who would choose any other field.

I am not pessimistic. Always, of course, there will be inventors who are really not technically competent and who delude themselves. But I believe that a real technical advance, the development of a good product that fills a need, or even a significant improvement of an established product will not go unrecognized. With a good product and a need for the product, someone, maybe even the inventor, will find a way to commercialize it.

9 . . . wherein all the factors necessary for a creative nation are present, and suggestions are made for methods to encourage creativity, and there is hope

Once upon a time there was a country, a leader among leaders in world finances, and science, and technology. Products emerged from its technical cornucopia in such abundance and with so high a quality that it was the source of most of the world's medical advances, that its technical products were sold around the world, and that its balance of payments was the envy of other less technically-advanced nations. But now the horn of plenty no longer spewed forth its bounty, and the country found itself buying products from other countries that had formerly been its customers. And it yearned for the thing that would return it to its former status, namely, a way to convert ideas to saleable products.

Once upon a time there was a company, progressive in its policies and successful in its financial endeavors. Success had come as the result of the development of a series of products coming from the fertile minds of its research people. But now the products were no longer forthcoming, and the company yearned for the only thing that it needed to keep its image of success intact, namely, new products.

Once upon a time there were inventors, keen of mind and ambitious for success. Ideas flowed from their fertile imaginations. They dreamed of the time when the value of their creations would be recognized, and they yearned for the only thing that they needed to bring their dreams to marketable products, namely, money.

The once upon a time might still be now. It seems even more like a fairy tale when we consider the fact that all of the elements necessary to satisfy everyone—the inventors, either private or institutional, the government, and industry—are available for bringing to fruition the desire of all parties involved.

Why has the supply of products decreased? Why has our balance of payments reversed? Why are there few new breakthroughs coming from industry? Why does the independent inventor, the greatest source of new ideas available, have to struggle for recognition? Why have the results of research creativity declined?

Certainly not because the people doing research today are any less creative than they were thirty years ago. Certainly, too, not because methods of doing research are less efficient. The development of sophisticated instrumentation is one of the great advances in science. There are instruments available today that make it possible to do in a few days what would have taken a year thirty years ago—if it could have been done at all. Certainly not because the store of basic knowledge is any less than it was thirty years ago. Certainly not because there is any less money available than there was in the early days.

If everything necessary for creativity is the same or better, why are the results worse? I think the simple answer is that people are not being allowed to create.

Why aren't they being allowed to create? From what I have said in previous chapters, I can relate a number of possible reasons.

Industrial creativity has declined, first, because government regulations have made it so difficult that it is now impossible for business to finance the research necessary to develop products and still fulfill its obligation to its stockholders.

Second, business leaders are not able to evaluate risk and they, therefore, take the easy way out by avoiding anything they can't prove to be safe.

Third, the image of what a successful business is has changed, and the requirements for success do not permit creative research.

Fourth, the self-image of business leaders has changed, so that they must operate a business to fulfill their own ambition, and this operation does not have room for creative research or, in fact, much creativity of any kind.

Fifth, the lack of appreciation in industry for any idea originating outside its own research prevents industry from taking advantage of a great source of creativity.

Money alone is not enough to solve the problem. But money used correctly is necessary to solve the problem. I think the problem can be solved. All the elements necessary for solutions are available. It is only necessary to bring them together.

Let's consider independent inventors. All they need is money. They have the ideas, but no one will give them the resources necessary to prove their ideas practical. The venture capitalist will not support their work because there is a perceived risk, and the venture capitalist does not wish to venture. The industrial laboratory will not support their work because of the barriers raised by the legal departments and because of the "Not Invented Here" syndrome present in practically all laboratories.

In the rare instance where inventors do get support from investors, what arrangements are made? They are rather simple. In exchange for financial support, the investors get as their return a certain percentage of any profits made from the project. The results of the work are patented. The inventors then commercialize the product either by licensing someone to do it under the patent, or, in some instances, using the patent to establish an entirely new company. The investor, the inventor, and the marketing company profit, and the general public has available a product it would not otherwise have.

These elements can be put together. Inventors with ideas need money, and are willing to give up a reasonable percentage of their future income in exchange for support. We have a government, desiring to stimulate research and already spending billions in an effort to do so, but unable to accomplish this because it cannot promise a return to the taxpayer on the money invested, and is afraid to risk the political criticism that might result from support without a specific and obvious financial return. We have an industry desiring to develop products from which most of the R&D risk has been removed.

How do we get off the merry-go-round? How do we break the cycle composed of the business that will not spend money on long-range problems, and the inventor who cannot get financing, and a government who cannot appear to be aiding business at the expense of the general public? It seems certain that the business administrator will not risk money on long-range projects. His or her immediate reputation is threatened if a company does not obtain a quick payback. It seems equally certain that the inventor outside industry will continue to be considered a risk. And politicians will continue to be politicians till the end of time. The business person wants products, the inventor wants to invent products. One key, then, would be to find a method to supply to business products from which the research risk had been removed with no cost to business.

I propose that the government act as a venture capital company to sponsor the research of independent inventors. The sponsorship should not be in the form of grants or contracts. The relationship of the inventors to the government would be exactly the same as would exist between an inventor and an investor. A contract would be signed giving the government a percentage of profits. The inventors could obtain patents, and commercialize their inventions any way they see fit.

Traditionally, venture capital investors want a return on their investment of 800% to 1,000% in three to five years. Obviously, the government should not expect this on all programs it sponsors. But suppose the government got back as a return only its original overall investment. Some projects would fail because the program would be designed to sponsor those things to which considerable risk was attached. Others would succeed. If the entire operation broke even, it would be a success. Inventors would have the right to commercialize their products in any way they chose. They could license them to an already existing organization, in which case the company would have a product without having had to invest in research and development. Or the inventors could arrange to manufacture and sell the product themselves.

The taxpayer would have lost nothing. On the contrary, totally aside from the sponsorship of research, there would be a tremendous benefit to the economy. Taxes would be paid by whomever commercializes the products. Industry would benefit. Possibly new businesses would result. Jobs would be created. New products would become available for export. There would be a mechanism by which creative individuals would test their ideas.

I am proposing that this type of government sponsorship be undertaken in no small way—not to the extent of a hundred thousand or a million dollars, but rather to the extent of a billion dollars. This is about 5% of the present total government sponsorship of research or, in other comparative terms, the cost of one battleship.

I can hear the reaction: another government handout. Asking the government to solve industry problems. Where there is government money there is government control. How can the government judge what project to support? It's not even a government problem.

I would be the last person in the world to suggest government support as a substitute for other support. Over the years I have objected vigorously, publicly and privately, to government grants to industry to support research. I had two reasons for objecting. Where the government supported research it did control the research and the results of the research, as witness the requirements that patents become the property of the government. My second reason is a more philosophic one. I don't believe an industrial organization is ever justified simply to substitute government money for its own money when engaging in a risk-taking activity like research. If a risk is justified the company should be willing to risk its own funds with the same confidence that it risks the taxpayers' dollars.

Should we expect that whatever government agency administers these venture funds will be able to select the proper projects for support? There are many people both in industry and in government who are capable of making such judgments. The difficulty lies in the implementation of action after judgment is made. If scientists in industry judge a project to be good, they also must judge when the company will receive its payback and how much the payback will be. If the criterion for acceptance is that the project must return a high percentage of the investment in three years, obviously many projects will be rejected. I would anticipate no such criterion for government support. The basis for deciding whether or not a project is practical and could become commercialized would be the same. However, the people making the decisions would not be influenced by the fact that they must report a profit every three months, nor would they be influenced by security analysts making predictions, nor would they have to suffer the indignities of having their peers make them a subject of derogatory cocktail party discussion. Their primary objective would be to make it possible for creative individuals who have no facilities of their own to prove or disprove their concepts. The return to the nation would be in the creation of new jobs, the availability of new products, the strengthening of the entire economy and, a not unimportant consideration, the building of an atmosphere that says that the creative individual is an important and recognized part of our national plan.

It is not difficult to rationalize the expending of national resources to restore the productivity of our technical effort, whether it be in industry or through independent inventors. Eventually everything becomes industry, because it is only through industrialization that an idea becomes a product. The commercialization can come about by the formation of an entirely new company, or it can result from an established organization taking over from the original developer.

Another way in which government could aid industry at no cost but with a potential for profit would be to liberalize the policy on granting exclusive licenses on patents already owned by the government. The concept of legal monopolies, patents, and copyrights, goes all the way back to the Constitution of the United States. Under the section on "Powers Vested in Congress" is one: "To promote the progress of science and useful arts by securing for a limited time to authors and inventors the exclusive right to their respective writings and inventions."

Abraham Lincoln, himself an inventor and the holder of a patent on a boat designed to float itself over obstacles, said: "The patent system added the fuel of interest to the fire of genius.' Thomas Jefferson said: "Issue of patents for new discoveries has given a spring to inventions beyond my conception."

Inventors are granted the exclusive right to use the results of the work they have patented. They are granted a legal monopoly. In exchange for that monopoly the inventors are obligated to make the results of their work public, not so that others can use that specific invention, but so that they might use the new knowledge as a base for further invention. The patent file is a tremendous source of data. For example, the Director of the Department of Commerce Office of Technology Assessment and Forecast reported in April of 1979 that the Patent System contained some 22.5 million documents of domestic and foreign technology distributed through some 100,000 subclassifications. He stated that 70.7% of the overall information in the files contains technology not disclosed anywhere else in the literature, and another 13.3% contains information only partially disclosed elsewhere!

The patent file is also a source of products. But it is not being used because the government is not willing to grant an exclusive license to a patent that it owns. Without such protection no company would be willing to invest in commercialization. Even an existing product demands thousands, and possibly millions, of investment dollars to bring it to the market. The product must be put into a marketable form, markets must be developed, sales and production organizations must be formed.

An indication of the requirement of exclusivity by industry is apparent in the fact that there are now some 30,000 patents owned by the government, but presently entirely unused. They are available, but only for non-exclusive licensing. In other words, as many companies as desire to take a license may do so. There is one hitch to this. Someone must do the work necessary to develop the invention, to make it practical, and then develop the market. This is expensive. No company is going to do this, realizing that when all the groundwork has been done, one or more competitors may enter the competition without having to duplicate the investment of the pioneering company.

It is a very short-sighted policy on the part of the government. The emotional appeal of the policy is evident. The government—or rather individual members of the bureaucracy—take the position that, since the patents resulted from the expenditure of taxpayers' money, all taxpayers should profit from them. It would, then, be unjust to allow only one company to profit, and thus exclude all other companies. It is a politically expedient position, because it makes it appear that the public is being protected from the voracious corporation. But the results have been that no one has benefited. The patents just sit there. The politicians are secure in their position of protecting the taxpayer, when actually they are preventing benefit to the taxpayer. It would seem that someone would recognize that it is better to get a royalty return from one company then to get nothing from all the companies.

It should be apparent that I am not proposing that the government act as the savior of individual companies. I am not proposing anything that would cost the taxpayer money. I am proposing that recognition be given to the fact that the problem is larger than the progress of an individual company, or the well-being of an individual inventor. It quite literally extends to the matter of national survival. In recent times we have seen the dollar devaluated to a frightening point. The basic cause has been the unprecedented negative balance of trade—the fact that we are buying more products from abroad than we are selling abroad. How can it be that the most highly developed industrial nation in the world suddenly becomes dependent on imports? Even worse, how can it be that a significant portion of this negative balance is based on importing products from abroad whose manufacture is based on technology developed here?

The simplistic reason is that, although our technological effort is still efficient, we have not used that technology to develop products for worldwide sales. Related to that is the fact that we, as a nation, have still not come to a full realization of the importance of technology for our survival, not as a military or political power, but as a financial power.

Our technology has not been used to develop products saleable around the world. Examples abound. For instance, our long-range aircraft are of no value in schedules that include even international travel in Europe. Consequently, short-range English, French, and German planes take the market. Our clothing manufacturers retain American styles and designs. They even attempt to sell American sizes in the Japanese market. I can well remember the warnings about the Japanese technical capability uttered years ago. But have no fear. We will allow the Japanese to make non-sophisticated equipment, things we don't want. We will stick to computers, tape recorders, and cameras.

Now we see foreign countries equalling or surpassing us both in capability and in effort. It is comforting to see that we are spending twice as much for research as Japan. But hold. Japan spends less than 10% of its budget on defense and space-related projects, the United States spends over 60%. Based on figures reported in the Science Indicators published by the National Science Foundation in 1976, Japan had three scientists working in research and development for every four in the United States, and the difference is decreasing. Defense spending adds nothing to our economic base, so our analysis should be based only on research and development that adds to our goods and services, or that improves the quality of life.

The rest of the world recognized the importance of our trade imbalance before we did, as witness the decline in the value of the dollar. It is easy to ask "How can our exports be so important when only about 5% of our GNP goes into exports?" Exports are crucially important, regardless of the amount, if we start importing appreciably more than we export. To use a wellworn phrase, we must redirect our research and development goals. It is easier said than done.

In the first place, an industrial organization has not only national goals, but primarily it has its own corporate goals. These goals are to survive and grow. When the company competes in the U.S., its competitors are playing by the same rules. When it competes abroad, the rules are changed. For example, other countries, recognizing the value of saleable products from technology, subsidize some parts of industry to allow them to bid on products in foreign markets. Even if we have superior technology in a field it is not possible to translate it into a product to compete in price with a competitive foreign product subsidized by a government.

In the political atmosphere in which we live it is almost unthinkable to consider such a situation. Even if politicians realized the importance of technological expansion as a major factor in our economy they probably would not have the courage to suggest it, because they would immediately be accused of being a tool of industry, interested only in fattening profits rather than preserving our national economy.

In a competition where the costs, and therefore the prices of each competitor, are increasing equally, the odds of success remain equal. Where the costs of one are going up, and those of another are not, the odds favor the company with the lower prices. The costs for all U.S. companies are going up and, therefore, competition with foreign industry becomes more unequal.

But it is not sufficient to say merely that costs are going up. Coupled with that must be the statement that production efficiency is decreasing. The final price of a product is not determinded only by the hourly rate of the worker or the price of raw material. It is based on the cost to produce each unit—the unit cost—of a product. It does not make any difference if wages increase as long as more units are produced for that increase. However, for about the last ten years productivity has been increasing very slowly. From about 1948 to 1968, industry and farms showed a productivity increase that averaged about 3.2% per year. According to the Bureau of Labor Statistics, since 1968 the gains have averaged only 1.6%, with the first quarter of 1978 showing a 3% *loss*. This loss is caused by a multitude of factors—the lack of the ability of plants to operate at capacity and so take advantage of economies of scale, an influx of inexperienced women and minorities into the labor force, higher energy costs and, probably most important, increasing government regulations.

Industry, too, could be more creative in investigating and developing total fields of technology with a very small initial investment. I can suggest one approach. For a commitment of probably no more than a million dollars, it would be possible, not only to determine what unpublished work is available, but to investigate the practicality of either concepts or developed prototypes in special areas. Funding new sources of energy is, or should be, one of our top national priorities. A number of specific problems can be isolated from the generality of "the energy problem."

Take the development of energy from fuel cells as an example. A fuel cell converts the energy generated in a chemical reaction directly into electricity. The basic understanding of fuel cell technology has been available for decades. Fuel cells have been used to power tractors and automobiles on an experimental basis. However, the process has not been economically feasible.

With such a broad base of data to draw from, I think I would be willing to gamble that, somewhere, there is an inventor who has the solution to the problem, but has not had the opportunity to prove his or her idea. How do we gain access to these ideas?

The first action would be to make it known to the inventors that there is someone who not only is interested in new ideas but who is willing to do something about encouraging their use. I would advertise for ideas in journals, or even newspapers read by inventors. The ad would read something like this:

WANTED: FUEL CELL INVENTORS

> We are a technology-based group with a proven track record for taking concepts to the product stage. We now want to establish ourselves as the clearinghouse for fuel cell technology. We guarantee a sympathetic hearing to all inventors. Should our evaluation indicate to us that your project is worth pursuing, we shall attempt to obtain financial support to prove feasibility and to take the eventual product or process to market.

The minimum expenditure for a group wanting to take this action would be the amount spent for the initial advertising. Should there be no worthwhile projects submitted, this would be the only cost involved, and it would be a total loss. However, should encouraging projects turn up, it would still be within the control of the sponsors to decide how much should be risked in their development. Thus, for a rather nominal capital risk for advertising, it would be possible to determine whether or not there really are worthwhile ideas "out there," waiting to be nourished.

The whole program is under the total control of the sponsor. If nothing develops, the initial investment is all that is lost. If encouraging ideas are discovered, the judgment of the sponsor is the deciding factor in how much is risked and for what benefit. I would bet there are inventors with good ideas who would welcome such an opportunity, and I would bet that some of the ideas could be commercialized at a profit.

I have mentioned the cooperation of industry and university research as a possible source of great advances in science and technology. In view of the obstacles that seem to keep them apart as stated in Chapter VIII, what can be done to stimulate cooperation? Academic research wants financial support, freedom to work on problems of its choice, freedom to publish. Industrial research also wants financial support, freedom to work on problems of its selection, and financial return. Unstated but nevertheless present in academic research is also the desire for financial reward.

There are many problems of common interest to both industry and academia. Therefore, it should not be difficult to find common projects that would satisfy the criterion of freedom of choice of a working area. Once a problem was agreed upon between a company and a university, a budget would be presented. The company would agree to pay half the expenses, with the government paying the other half to the university. This would fulfill the requirements of funding for the university and cut in half the project cost to the company. The company and the university would agree on a publication and patent policy. If patented results came from the projects, total rights would remain with the company and the university, with the government receiving as its return a small royalty on anything that is commercialized. This would give the possibility of financial return to both parties to the venture, and would also allow the government some chance to have its investment returned. If the project would be in a basic research area, the company would have an advantage in the use of the results, since the company would have the information before it became public knowledge. Eventually, however, knowledge would become public and available for free use, as a result of the government participation.

I cannot see why there should be any objection to this from those distributing government grants even if some of the present grant money would be used for this purpose. There is no discernible government plan for the effective use of research money now. With joint consideration from both academic research and industrial research it is probable that better analyses and selection of research projects would be carried out than is the case at the present time.

In discussing the factors contributing to the decline of industrial creativity I have touched on overcautious business leaders, too much regulation, not enough recognition of independent inventors, lack of cooperation between academia and industry, and the government patent policy. In suggesting possible solutions to these problems I have, perhaps, revealed the one factor to which I do not believe there is any solution. I do not believe the cautious business leader can ever become a risk-taking leader, even when that term is meant to reflect a prudent risk-taking leader. There are in our industrial world today possibly a half-dozen leaders of major companies who are willing to take a long-term risk for a major gain. The reasons for caution I have already explained. It is necessary for every leader to succeed. One of the basic measures of success is continued and continuously higher profits, measured even on a quarterly basis. When a project fails, it reduces the profits. If a project does not pay off for five or ten years, it reduces immediate profits.

What I have attempted to do might appear simplistic. I have said there could be ways to eliminate the necessity for company leaders to be creative, or to evaluate creativity in technical fields. The area of creativity has been removed from management. The least predictable and the most hazardous part of any original project is that involved in proving that a concept is practical. Can a particular type of solar cell convert the sun's energy to electricity? Will a particular chemical cure a disease? Can a synthetic substitute for gasoline be made?

The decision as to whether or not to spend millions of dollars and years of work to see whether or not a concept is valid is an entirely different order of magnitude of risk than the decision to develop or not develop an already existing product. It is this belief that has led me to emphasize the importance of encouraging invention outside industry. When administrators see finished products, the most dangerous part of their risk—that of research and development—has been removed. The remaining risk is one they feel comfortable with—that of deciding whether or not a product can be sold.

In a similar vein, the administrator might be tempted to risk if risk didn't cost anything—that is, if noncompany money were used. If the project failed, profits would not suffer. If it succeeded, he or she would look like a real hero. Thus, the reason for tax incentives and matching grants. I think it is perfectly reasonable to suggest using tax money for such incentives. I think there is no alternative. The economic future of the nation is at stake, and if industry cannot or will not take the initiative in finding a solution then it must be prodded into taking action. Observe that this is no windfall for industry. This plan does not substitute tax money for money that industry would have spent anyway. It is a way of getting done what would not be done any other way, it offers the potential for great reward for the nation both in creative advances and the establishment of new jobs, it does not give industry something for nothing, and it even provides for the possible return to the government of the money it has invested.

I have expressed before my strong feelings about government money being used by industry, if the money simply replaces that which industry would have spent as part of its own program. There are some projects, vital to the well-being of the nation, that are too big for a single company to undertake. For those projects I would propose tax credits to stimulate entrance into the field. The company would be taking some risk, but it would also reap great benefits from success. New forms of energy would be one such field that would qualify for credits.

And I would suggest more lenient antitrust attitudes toward those companies that might want to engage in cooperative work on a major problem. However, I cannot suggest many problems that would benefit from such cooperation. In general, I believe that the *competition* of two organizations is more likely to result in successful solutions than is *cooperation*. I should emphasize that this belief applies only to those situations in which the solution to a problem is being sought. The more different approaches that are brought to bear, the more chance there is that one will be successful. I do believe there is room for cooperation once a solution has been determined and it is then only necessary to carry out the required development work. One of the benefits of changing this interpretation of the laws would be a tacit recognition by government that industry is vital to the future of the country, that government should be the stimulator of industry rather than its brake. One of the reasons given for the rapid rise of Japanese industry has been the attitude of its government. There, industry and government are partners, not antagonists. On the minus side of such an atmosphere is the lack of real creativity in Japanese industry. Without competition, but with cooperation between companies, operation is much more efficient. But it results in efficient imitation, not efficient originality. I speak only for a balance between our own repressive system and a system almost totally permissive in its control of competition.

Whether it is the product of the artist, the artisan, the business executive, or the scientist, the final test of whether creativity has served its highest purpose is not that it adds to our comfort or our power or our wealth. It is whether or not it has added to our quality of life, our opportunity to live with added dignity, our sense of truth and beauty, our appreciation of things of the spirit. Industrial creativity might be considered rather limited in its scope. How can the results of industry increase our sense of truth and beauty?

There is a grandeur in any act of creativity. There is mystery in the formation of a new idea. A new idea is one of the most precious things in the world. It comes along all too seldom. But when it does, revolutions are caused, industries are born and, in some cases, other industries die as a result of the same idea. It is a marvelous thing to know you have thought something no one else has ever thought, and done something no one else has ever done.

The creativity of the artist or that of the industrialist, whether it be administrator or scientist, are the same and yet different. When an artist paints a masterpiece it is his or her painting forevermore. Other artists might paint other masterpieces, but his or hers will not be displaced. The painting might be sold, but it still remains always the creation of the artist.

In industry a breakthrough brings the same satisfaction to its originator as does the painting to the artist. But quite often it is only temporary. The advance will always be remembered as an advance, but contrary to the permanence of a painted picture, the result might be superceded or even overturned by tomorrow's advance. It is one of the requirements of science.

Our country was founded by creative people, and it grew and prospered as a result of creative ideas. Our predecessors were imaginative and innovative. And they were risk takers. Results of their vision have allowed us to increase the quality of life over what they had known. Theirs was a vision that allows us to say that we now, possibly as the first generation in history to do so, are able to devote more of our time to improving the human race than in working only to remain alive.

Business is the greatest influence for good of any organization in the world. I said that earlier. But, particularly where creativity is involved, business can also be the greatest inhibitor. Ideas in the business world are of value only when they are developed into practicalities. I have pointed out several segments of the country where ideas are generated—universities, individuals, government laboratories, and in industry itself. The works of the artist, the sculptor, the poet, the philosopher—those things that contribute only to our intellectual or esthetic well-being, do not need industry. But the works of the economist and the scientist and the engineer require development. An industrial organization is the vital mechanism for developing an idea into a product. It is both a conduit and a filter. The filter can eliminate all ideas, or it can eliminate any proportion of ideas, good or bad. In recent years the pores of the filter have become so small that more and more ideas are being rejected. But larger filters are available, if our leaders would but seek them out.

The United States, in spite of its decline, is still the leading industrial nation in the world. It is the most creative nation in the world. It has the greatest potential for good of any nation in the world. How is that potential fulfilled?

Thomas Wolfe said:

> I think the enemy is here before us. I think the enemy is simply selfishness and compulsive greed. Go, seeker, if you will, throughout the land, and you will find us burning in the night.
>
> To every man his chance—to every man, regardless of his birth, his shining golden opportunity—to every man the right to live, to work, to be himself, and to become whatever thing his mankind and his vision can make him . . . this is the promise of America.[1]

Business, by creating jobs, by raising the economic standard, has played a great role in allowing every person to have his or her chance, his or her shining golden opportunity. Business, too, has had its own chance and opportunity to be born, to grow, to thrive. It has grown because creative people in government allowed it to grow, and because creative people in industry directed its growth, and because ours was a land where creativity and imagination and courage and daring were appreciated and almost demanded.

I don't think the character of our people has changed that

much in a generation or two generations. Perhaps we shall once again be blessed with a government that regards creative endeavors as being important enough to justify political criticism for their support. Perhaps we shall once again have a citizenry that regards risk as a necessity for biological existence. Perhaps we shall once again see a generation of industrial leaders who recognize and admit that a company exists not just to make a profit but to serve humanity.

Perhaps.

notes

Chapter One

1. *The Wall Street Journal,* 10 October 1978.
2. *Short Stories of Saki* (New York: Viking Press, 1930), p. 601.

Chapter Two

1. Yevgeny Zamiatin, *The Dragon* (New York: Random House, 1967).
2. Conference Board Report 779, *Annual Survey of Corporate Contributions,* 1980 ed.
3. "The Growing Impact of Business Giving," *The Nation's Business,* October 1980, p. 67.
4. Milton Friedman, *Capitalism and Freedom* (Chicago: University of Chicago Press, 1962).
5. Adam Smith, *An Inquiry into the Nature and Cause of the Wealth of Nations,* R. H. Campbell and A. S. Skinner, eds. (Oxford: Clarendon Press, 1976).
6. Ibid.

Chapter Three

1. William E. Simon, *A Time for Truth* (New York: Reader's Digest Press, 1978).
2. *Dun's Review,* December 1979.
3. *Nation's Business,* October 1978.
4. Congressman Alan Strangeland, personal communication to author.
5. *Saturday Review,* 20 January 1979.
6. *The Nation's Business,* October 1978.
7. Paul McAvoy, quoted in *The Nation's Business,* June 1978, p. 19.

8. World Health Organization Expert Committee on the Prevention of Cancer, TECH. REP. SER. 276, *WHO Chron.* 18 (1964) 323.

9. "Cancer and the Environment—Higginson Speaks Out," *Science* 205 (1979) 1363.

10. John Cairns, *Cancer, Science, and Society* (San Francisco: W. H. Freeman & Co., 1978).

11. *Chemical Week,* 19 December 1979, p. 70; *Chemical and Engineering News,* 24 December 1979, p. 13.

12. *Chemical Week,* 12 January 1980, p. 35.

13. A. S. Morrison and J. E. Buring in *New England Journal of Medicine* 302 (1980) 537.

14. E. L. Wynder and S. D. Stillman in *Science* 207 (1980) 1214.

15. FDA Contract 74-2181, 18 May 1978.

16. *The Sciences,* January 1980, p. 7.

17. U. S. Dept. of Commerce, Food Safety Policy Scientific and Societal Consideration, Part 2, U. S. Dept. of Commerce Publication PB-292-069, 1 March 1979.

Chapter Four

1. James Russell Lowell, *A Fable for Critics* (Cambridge, Massachusetts: The Riverside Press: Houghton, Mifflin and Co., 1890).

2. Report of the President's Commission on The Accident at Three Mile Island, *The Need for Change: The Legacy of TMI,* Washington, D.C., October 1979.

3. *The Wall Street Journal,* 5 June 1979.

4. Report AECB 1119 Atomic Energy Control Board, Risk of Energy Production, Canada, March 1978, 2nd ed. May 1978, 3rd ed. Nov. 1978.

5. Chemical Abstracts Service, Columbus, Ohio, personal communication to author.

6. Marsh and McLennan Co., Inc., "Risk in a Complex Society," 1980.

Chapter Five

1. *The Nation's Business,* June 1979, p. 83.

2. American Association for the Advancement of Science, *R&D, Industry and the Economy,* Research and Development: AAAS Report III (Washington, D.C.).

Chapter Six

1. *Science Indicators 1974,* Report of the National Science Board, Washington, D.C., 1975.

2. American Association for the Advancement of Science, *R&D, Industry and the Economy,* Research and Development: AAAS Report III (Washington, D.C.) p. 53.

3. National Science Foundation, *National Patterns of R&D Resources, Funds and Manpower in the U.S. 1953-1977,* N.S.F. 77-310 (Washington, D.C.).

4. *Chemical and Engineering News,* 3 October 1977.

5. *The Wall Street Journal,* 18 October 1977.

6. Ibid.

7. Howard J. Sanders, "Recognition for Employed Inventors, Special Report," *Chemical and Engineering News,* 26 May 1980, p. 32.

8. Ronald W. Hansen, *The Pharmaceutical Development Process: Estimate of Current Development Costs and Times and the Effects of Regulatory Changes* (Rochester, N.Y.: University of Rochester, 1977).

9. W. Wardell and L. Lasagna, *Regulation and Drug Development* (Washington, D.C.: American Enterprise Institute, 1975).

10. *Fact Book: 1980* (Washington, D.C.: Pharmaceutical Manufacturers Association, 1980) p. 23.

11. William Wardell, "The Rate of Development of New Drugs in the U.S. 1963 through 1975," *Clinical Pharmacology and Therapeutics,* 24 (1978) 153.

12. American Association for the Advancement of Science, *R&D, Industry and the Economy,* Research and Development: AAAS Report III (Washington, D.C.) p. 56.

Chapter Seven

1. *Science and Technology Policy: Perspective for the 1980s,* Annals of the N.Y. Academy of Science 334 (1979) 158.

2. Office of Management and Budget Ad Hoc, Interagency Panel, *Small Firms and Federal Research and Development, A Report to the Office of Federal Procurement Policy,* 24 February 1977.

3. U.S. Department of Commerce Panel on Invention and Innovation, *Technology Innovation, Its Environment and Management,* (Washington, D.C.)

Chapter Eight

1. James Boswell, *Life of Samuel Johnson* (New York: The Modern Library, 1931), p. 965.
2. Elting E. Morison, *Men, Machines and Modern Times* (Cambridge, Mass.: MIT Press, 1966), p. 147.
3. Ibid, p. 171.

Chapter Nine

1. Thomas Wolfe, *You Can't Go Home Again* (New York: Harper Torchbooks, 1940).

index